Active Geometry

David A. Thomas

Montana State University

Brooks/Cole Publishing Company

I(T)P® *An International Thomson Publishing Company*

Pacific Grove • Albany • Belmont • Bonn • Boston • Cincinnati • Detroit • Johannesburg
London • Madrid • Melbourne • Mexico City• New York • Paris • Singapore
Tokyo • Toronto • Washington

GWO A GARY W. OSTEDT BOOK

Publisher: *Gary W. Ostedt*
Editorial Assistant: *Carol Ann Benedict*
Marketing Team : *Caroline Croley and Michele Mootz*
Production Editor: *Mary Vezilich*
Cover Design: *Bob Western*
Printing and Binding: *Patterson Printing*

For more information, contact:

BROOKS/COLE PUBLISHING
511 Forest Lodge Road
Pacific Grove, CA 93950
USA

International Thomson Editores
Seneca 53
Col. Polanco
11560 México, D. F., México

International Thomson Publishing Europe
Berkshire House 168-173
High Holborn
London WC1V 7AA
England

International Thomson Publishing Japan
Hirakawacho Kyowa Building, 3F 418
2-2-1 Hirakawacho
Chiyoda-ku, Tokyo 102
Japan

Thomas Nelson Australia
102 Dodds Street
South Melbourne, 3205
Victoria, Australia

International Thomson Publishing Asia
221 Henderson Road
#05-10 Henderson Building
Singapore 0315

Nelson Canada
1120 Birchmount Road
Scarborough, Ontario
Canada M1K 5G4

International Thomson Publishing GmbH
Königswinterer Strasse
53227 Bonn
Germany

Printed in the United States of America

10 9 8 7 6 5 4 3 2 1

Library of Congress Cataloging-in-Publication Data

Thomas, David A. (David Allen) , [date]
    Active Geometry / David A. Thomas.
       p.   cm.
    Includes bibliographical references.
    ISBN 0-543-34485-2 (pbk.)
    1. Geometry—Computer-assisted instruction.   2. Geometry—Study
and teaching (Secondary) I. Title.
QA462.2.C65T48 1997
516'.04 ' 078553--dc21
                                   97-45599
                                     CIP

To Cynthia

Acknowledgements

The author wishes to acknowledge the following individuals for their encouragement, insightful suggestions, and creative contributions to this work: James Smart, Robert Fixen, Brian Beaudrie, Paul Rozman, Donald Harder, Mark Plante, Timothy Hall, Robert Blakeman, Joseph Richards, Virginia Hammond, Stephanie Stevenson, Krista Johnson, Kipp Lewis, Jenny Wickum, June Skillingberg, and Cynthia Thomas.

PREFACE

Active Geometry was written for use in "geometry for teachers" courses at Montana State University – Bozeman, MT. Consistent with recommendations of the National Council of Teachers of Mathematics and other professional organizations engaged in the reform of school mathematics, *Active Geometry* introduces students to both Euclidean and non-Euclidean geometries, emphasizing the student's role as investigator, problem solver, and mathematical communicator.

Directed activities begin by focusing student attention on a mathematical theme or topic, then move quickly to hands-on investigations using computer modeling and analysis tools and data files selected and/or created by the author. Typically, students take measurements or perform transformations on a geometrical model. Their objective is to formulate conjectures concerning the general features of the model, test their conjectures, identify invariant points and lines under various transformations, and verbalize what they see and think. Through this process, students simultaneously develop mathematical insight, mathematical language, and confidence in their own ability to discover and create mathematical knowledge, given reasonable support.

Teaching students how to participate in the discovery of mathematics is the whole purpose of *Active Geometry*. As a result, *Active Geometry* is inductive by design and highly selective in its content. No attempt is made to systematically organize and present all that students might need to know about geometry. Books such as James R. Smart's *Modern Geometries, Fifth Edition* (Brooks/Cole Publishing Company) do that job admirably. *Active Geometry* is best conceived of as a supplement to such texts.

Unlike most books, *Active Geometry* is not complete in itself. All necessary computer software and data files are available at no cost at the Brooks/Cole World Wide Web (WWW) site. Links to the latest versions of The Geometers Sketchpad, NIH. Scion Image, Netscape Navigator, Microsoft Internet Explorer, and other tools make it easy for users with Internet access to obtain the software tools specified in *Active Geometry*. All data files used in *Active Geometry* are also maintained at the same site.

While *Active Geometry* began as a supplement for a university level geometry course for teachers, it is the author's hope that these activities will find their way into high school geometry courses. The reform of school mathematics must assert itself at the K-12 classroom level, impacting the lives of both teachers and students. To achieve this end, teachers need classroom materials that engage their students' curiosity and imagination. *Active Geometry* provides students with an opportunity to learn geometry their way, through investigation and dialogue.

Comments and suggestions are appreciated.

David A. Thomas

Correlation Table

Chapter Titles in James R. Smart, *Modern Geometries, Fifth Edition*
1 Sets of Axioms and Finite Geometries
2 Geometric Transformations
3 Convexity
4 Modern Euclidean Geometry, Theory and Application
5 Constructions
6 The Transformation of Inversion
7 Projective Geometry
8 Introduction to Topological Transformations
9 Non-Euclidean Geometries

Page in Thomas	Section in Smart	Page in Thomas	Section in Smart
2	4.1	62	4.6
4	4.1	63	4.6
6	4.1	64	4.6
8	4.1	72	6.1
10	4.1	73	6.1
11	4.1	74	6.1
12	4.1	75	6.2
13	4.3	76	6.1
14	4.3	77	6.3
15	4.3	80	7.1
17	4.3	82	7.1
19	3.2	83	7.1
20	3.6	84	7.1
21	6.2	85	7.1
22	6.2	86	7.1
23	6.2 & 7.4	88	7.1
26	7.4 & 7.7	89	7.7
27	7.4 & 77	90	7.7
30	9.1 - 9.4	91	7.7
33	9.1 - 9.4	103	2.9
35	9.1 - 9.4	104	2.9
37	1.2	105	2.9
38	1.3	106	2.9
39	1.5	107	2.9
41	1.5	108	2.9
43	2.3	109	2.9
45	2.3	110	2.9
46	2.3	113	2.9
47	2.3	115	2.9
51	2.4 & 2.5	118	2.9
59	4.6	132	2.9
60	4.6	135	8.4
61	4.6	137	8.2

CONTENTS

CHAPTER 1

CIRCLES, TRIANGLES, AND SEGMENTS IN EUCLIDEAN GEOMETRY

The great city of Alexandria, Egypt, was founded in 332 B.C. by Alexander the Great. Located at the intersection of important trade routes, the city also became a crossroads of learning. Ptolemy began building the University of Alexandria, the world's first great university, in 306 B.C. and opened the doors of the institution about 300 B.C.. A Greek mathematician named Euclid was chosen to head the department of mathematics. Among his many works, Euclid's *Elements* is the best known. For over two thousand years, the *Elements* has provided both teachers and learners a structured approach to geometry. Indeed, after the Bible, the *Elements* has probably been studied more widely and extensively than any other work.

The *Elements* consists of thirteen books, including 465 propositions or theorems. During this century, traditional courses in high school geometry have addressed much of the material found in Books 1, 3, 4, 6, 9, and 12. To millions of students whose schooling limited their geometrical training to these materials, Euclidean geometry is the only geometry. One of the goals of *Active Geometry* is to dispel that misconception in a way that builds student appreciation for and understanding of Euclidean geometry while motivating them to investigate other geometries and their applications in modern mathematics, science, and technology. Using powerful, interactive computer tools, students study geometry as explorers and discoverers in their own right. To provide structure for this sort of activity, directed inquiries emphasize the student's role as observer, thinker, and mathematical communicator. *Active Geometry* activities are designed to foster curiosity, structure exploration, develop confidence, and encourage students to formulate their own conjectures and concepts. On the other hand, no attempt is made to systematically present Euclidean geometry or any other geometry. Instead, *Active Geometry* is intended for use as a technology-based supplement in a geometry course for K-12 mathematics teachers.

The computer tool used most extensively in these activities is a dynamic modeling environment called the Geometers Sketchpad (GSP). Using the GSP, students conduct investigations, discover geometric relationships, and state the relationships as conjectures. In general, students discover these relationships by determining which geometric properties of a given construction are invariant under a wide range of modifications and transformations made easy by the GSP. Once a relationship is discovered, it is stated as a conjecture using standard mathematical language and notations. Students then search for the underlying mathematical reasons for the relationships they discover. This is what it means to be a geometer .

CIRCUMCENTER

Tool(s) Geometers Sketchpad   Data File(s)   gsp1   (Mac & PC)

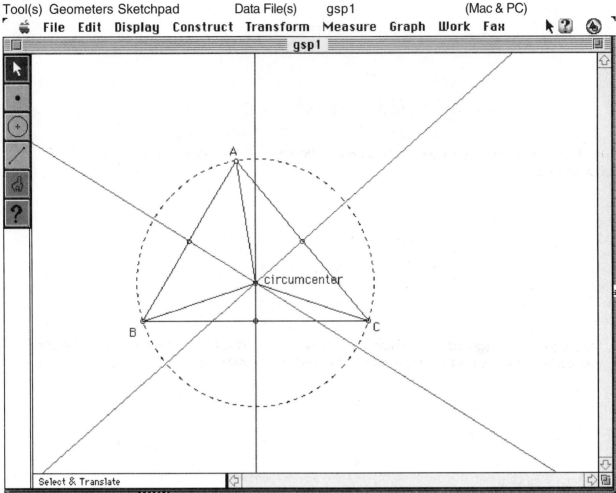

Focus

The intersection of the perpendicular bisectors of the sides of a triangle is called the circumcenter. The circle passing through the vertices of the triangle is called the circumcircle. The circumcenter is the center of the circumcircle.

Tasks

1. Using the GSP's arrow icon, drag the vertices around. Is it is possible for the circumcenter to lie on a vertex of the triangle? If so, under what circumstances? If not, why not?

2. Is it is possible for the circumcenter to lie on a side of the triangle? If so, under what circumstances? If not, why not?

3. Is it is possible for the circumcenter to lie outside of the triangle? If so, under what circumstances? If not, why not?

4. State a conjecture regarding the location of the circumcenter based on your observations. To prove your conjecture false, what sort of counter example would be necessary?

5. If the triangle is equilateral, where is the circumcenter?

Tool(s) Geometers Sketchpad   Data File(s)   gsp2   (Mac & PC)

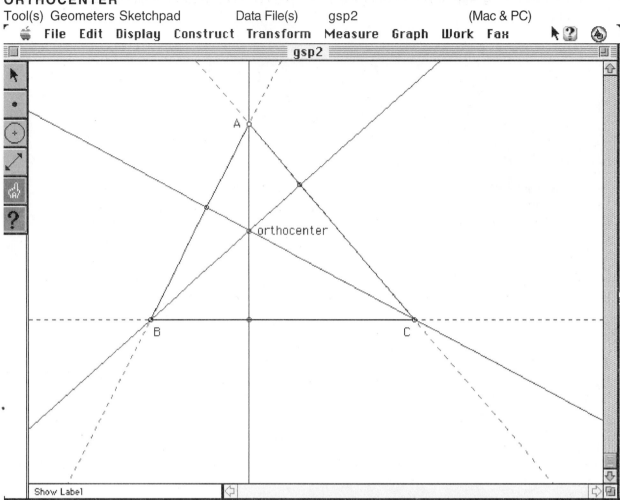

Focus

The point of intersection of the altitudes of a triangle is called the orthocenter.

Tasks

1. Using the GSP's arrow icon, drag the vertices around. Is it is possible for the orthocenter to lie on a vertex of the triangle? If so, under what circumstances? If not, why not?

2. Is it is possible for the orthocenter to lie on a side of the triangle? If so, under what circumstances? If not, why not?

3. Is it is possible for the orthocenter to lie outside of the triangle? If so, under what circumstances? If not, why not?

4. State a conjecture concerning possible locations of the orthocenter. To prove your conjecture false, what sort of counter example would be necessary?

5. If the triangle is equilateral, where is the orthocenter?

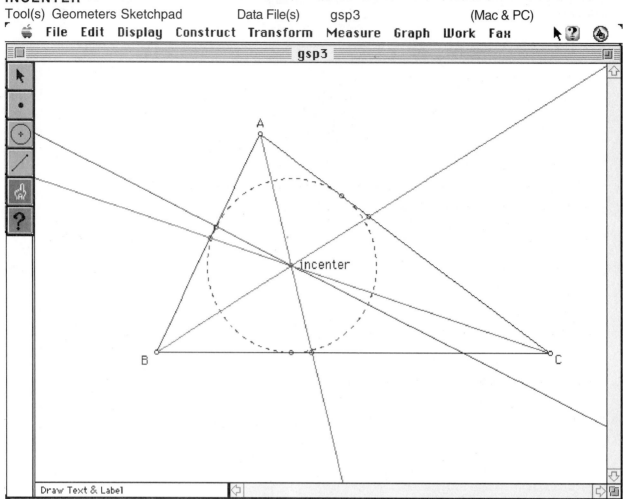

Focus

The internal bisectors of the angles of a triangle intersect at a point called the incenter. The circle tangent to the sides of the triangle is called the incircle. The center of the incircle is the incenter.

Tasks

1. Using the GSP's arrow icon, drag the vertices around. Is it is possible for the incenter to lie on a vertex of the triangle? If so, under what circumstances? If not, why not?

2. Is it is possible for the incenter to lie on a side of the triangle? If so, under what circumstances? If not, why not?

3. Is it is possible for the incenter to lie outside of the triangle? If so, under what circumstances? If not, why not?

4. State a conjecture concerning possible locations of the incenter. To prove your conjecture false, what sort of counter example would be necessary?

5. If the triangle is equilateral, where is the incenter?

6. Is it possible for the circumcenter, orthocenter, and incenter to coincide? If so, under what circumstances? If not, why not?

CENTROID

Tool(s) Geometers Sketchpad   Data File(s)   gsp4   (Mac & PC)

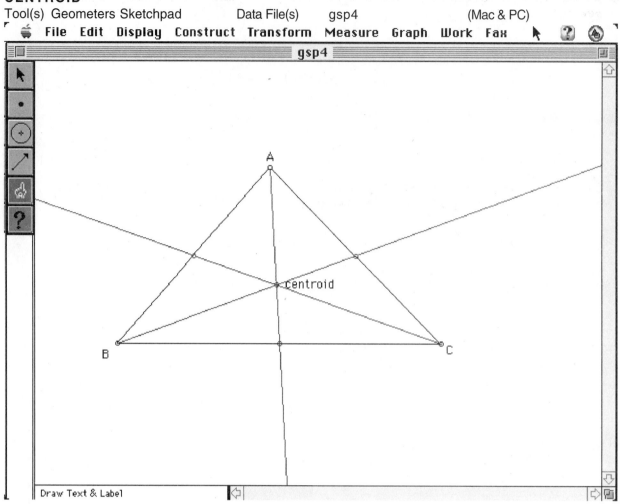

Focus

The medians of a triangle intersect at a point called the centroid. The centroid is also called the center of gravity of the triangle.

Tasks

1. Is it is possible for the centroid to lie on a vertex of the triangle? If so, under what circumstances? If not, why not?

2. Is it is possible for the centroid to lie on a side of the triangle? If so, under what circumstances? If not, why not?

3. Is it is possible for the centroid to lie outside of the triangle? If so, under what circumstances? If not, why not?

4. State a conjecture concerning possible locations of the centroid. To prove your conjecture false, what sort of counter example would be necessary?

5. Construct three polygon interiors, each of which is a triangle with one vertex at the centroid and two vertices on the original triangle. Measure and compare the areas of these triangles.

6. Using the GSP's arrow icon, drag the vertices around. State a conjecture based on your observations. To prove your conjecture false, what sort of counter example would be necessary?

7. Why do you think the centroid is called the center of gravity of the triangle?

SECANT AND TANGENT RELATIONSHIP

Tool(s) Geometers Sketchpad   Data File(s)   gsp5   (Mac & PC)

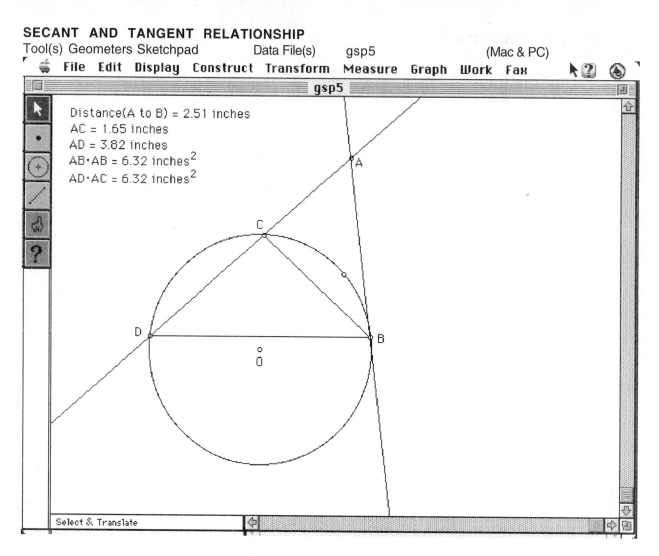

Focus

Investigate the relationship between the lengths of the segments drawn from an exterior point to the points of intersection of a secant with a circle and the length of the tangent from the point to the circle.

Tasks

1. Using the GSP's arrow icon, drag the vertices around and observe the measurements and calculations. State a conjecture based on your observations. To prove your conjecture false, what sort of counter example would be necessary?

2. Find two similar triangles in the figure and determine the ratio of their corresponding sides. Construct their polygon interiors and determine the ratio of their areas. How is the ratio of their areas related to the ratio of their sides?

PARTITION OF A SEGMENT

Tool(s) Geometers Sketchpad   Data File(s)   gsp6   (Mac & PC)

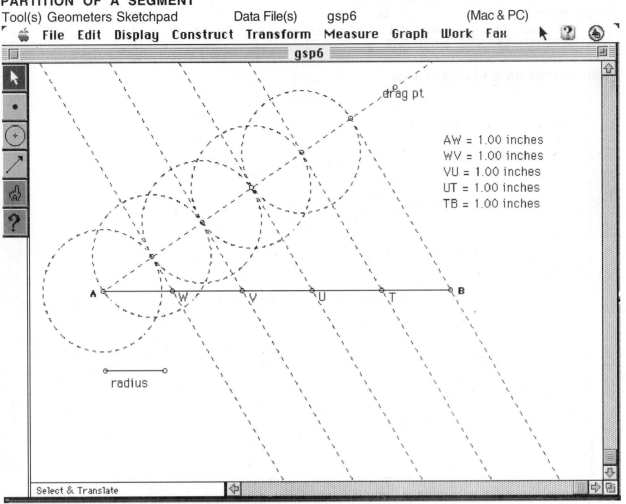

Focus

    A segment may be divided into any number of congruent parts.

Tasks

    1. Write a paragraph discussing the geometric basis for this construction.

    2. Describe at least one practical application of this procedure.

Tool(s) Geometers Sketchpad   Data File(s)   gsp7   (Mac & PC)

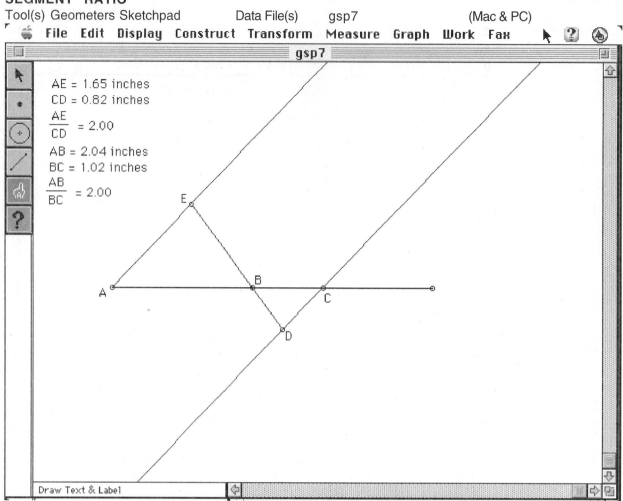

Focus

      A segment may be divided into two unequal but related parts.

Tasks

1. Using the GSP's arrow icon, drag points E and D around and observe the measurements and calculations. What do you notice about the ratios AE/CD and AB/BC? State a conjecture based on your observations. To prove your conjecture false, what sort of counter example would be necessary?

2. Write a paragraph discussing the geometric basis for this procedure and describing a practical application.

CONSTRUCTIONS

NINE-POINT CIRCLE
Tool(s) Geometers Sketchpad   Data File(s)   gsp8   (Mac & PC)

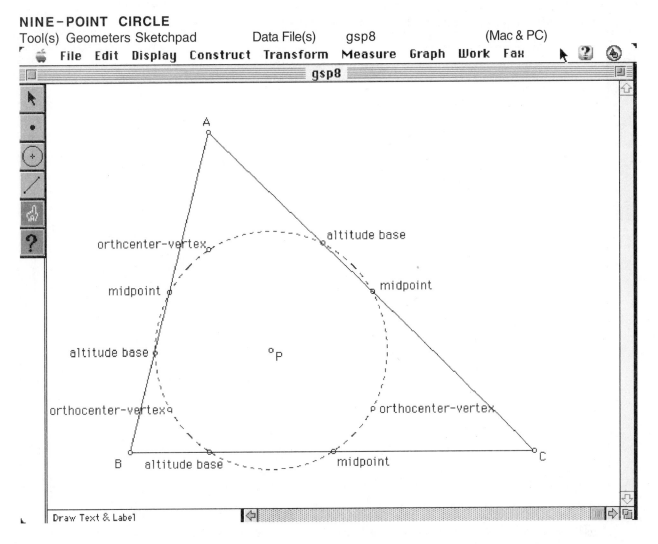

Focus
The midpoints of the sides of a triangle, the points of intersection of the altitudes and the sides, and the midpoints of the segments joining the orthocenter and the vertices of a triangle all lie on the nine-point circle. The nine-point circle was discovered by Karl Wilhelm Feurbach (1800 – 1834).

Tasks
1. Using the GSP's arrow icon, drag the vertices around and observe the effect on the nine-point circle. Is it possible to position the vertices so that each side is intersected only once by the circle? If so, under what circumstances does this occur? If not, why not?

2. When the triangle is isosceles, what happens to the nine points?

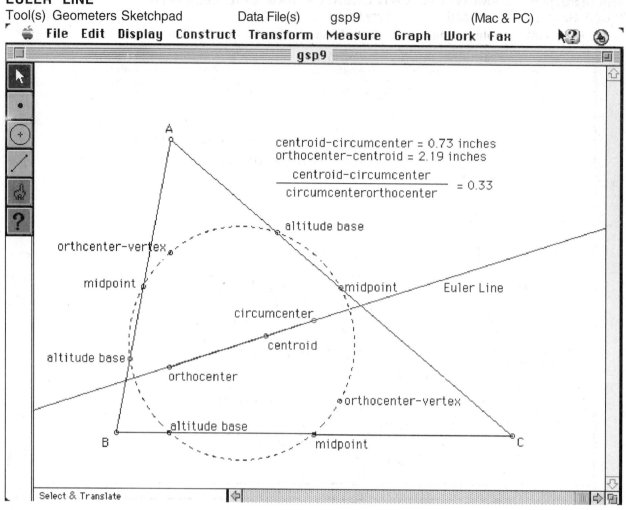

Focus

The line containing the orthocenter, centroid, and circumcenter is called the Euler Line, after Leonhard Euler (1707–1783).

Tasks

1. State a conjecture concerning the positions of the orthocenter, centroid, and circumcenter. To prove your conjecture false, what sort of counter example would be necessary?

2. State a conjecture concerning the following ratios: (centroid-center of the nine-point circle)/(centroid-circumcenter) and (orthocenter-center of the nine-point circle)/(orthocenter-circumcenter). To prove your conjecture false, what sort of counter example would be necessary?

THE SEGMENT JOINING THE ORTHOCENTER AND CIRCUMCENTER

Tool(s) Geometers Sketchpad   Data File(s)   gsp10   (Mac & PC)

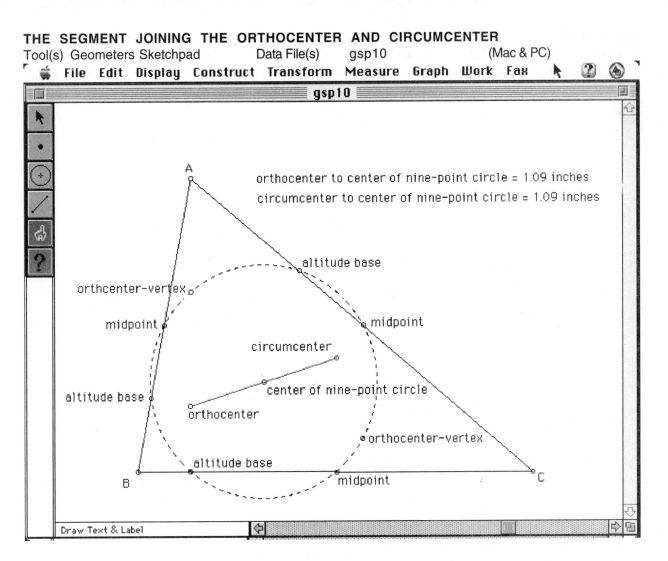

Focus

The relative positions of the orthocenter, circumcenter, and center of the nine-point circle.

Tasks

1. State a conjecture concerning the relative positions of the orthocenter, circumcenter, and center of the nine-point circle. To prove your conjecture false, what sort of counter example would be necessary?

2. Is it possible to position the vertices so that the orthocenter, circumcenter, and center of the nine-point circle coincide? If so, under what circumstances? If not, why not?

3. State a conjecture concerning the radius of the circumcircle and the radius of the nine – point circle To prove your conjecture false, what sort of counter example would be necessary?

4. State a conjecture concerning the manner in which the nine-point circle divides segments drawn from the orthocenter to points on the circumcircle. To prove your conjecture false, what sort of counter example would be necessary?

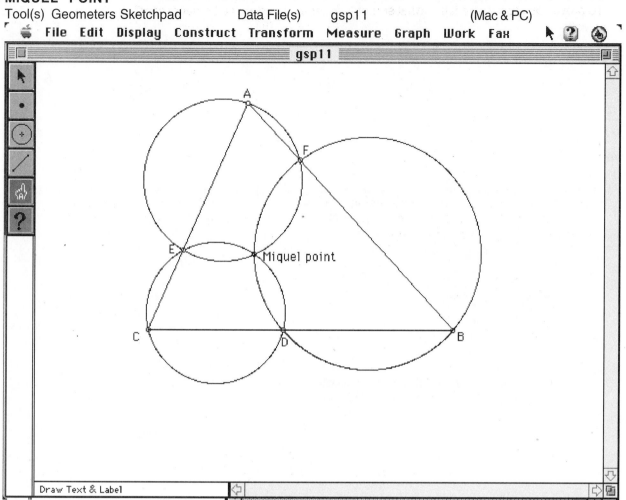

Focus

Any three noncollinear points determine a circle. In the figure above, each circle is determined by a vertex and a randomly positioned point on each adjacent side. These circles are concurrent at a point named after Auguste Miquel, who published a paper on the subject in 1838.

Tasks

1. Is it possible to position the vertices so that the Miquel point is on a vertex of the triangle? If so, under what circumstances? If not, why not?

2. Is it possible to position the vertices so that the Miquel point is on a side of the triangle? If so, under what circumstances? If not, why not?

3. Is it possible to position the vertices so that the Miquel point is outside the triangle? If so, under what circumstances? If not, why not?

4. Is it possible to position the vertices so that the Miquel point is on the circumcircle? If so, under what circumstances? If not, why not?

5. State a conjecture concerning the possible locations of the Miquel point. To prove your conjecture false, what sort of counter example would be necessary?

TANGENTS AND INSCRIBED POLYGONS

Tool(s) Geometers Sketchpad   Data File(s)   gsp12   (Mac & PC)

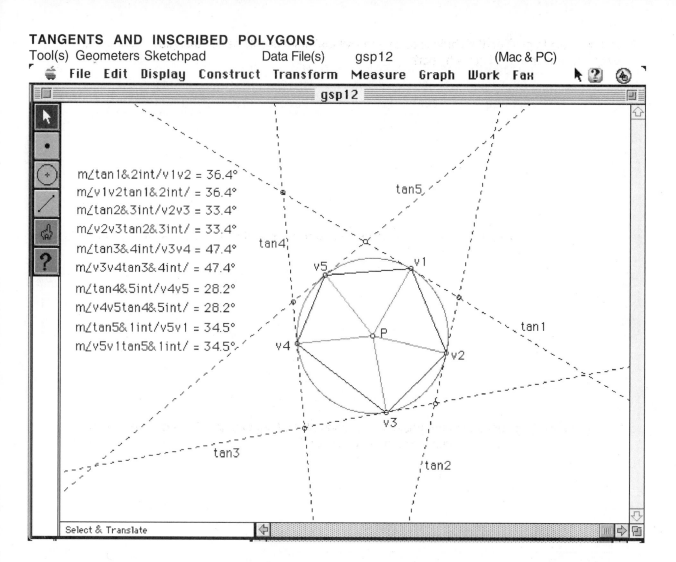

Focus

Tangents drawn at the vertices of an inscribed convex polygon and the angles they form exterior to the polygon and interior to the circle.

Tasks

1. After highlighting the angle measurements, use the Calculate option under the Measure menu to sum the angles. State a conjecture concerning the sum of the angles exterior to the inscribed polygon and interior to the circle. To prove your conjecture false, what sort of counter example would be necessary?

2. Construct the polygon interior and measure its area and perimeter. Do either of these quantities remain constant when a vertex is moved?

TANGENTS AND CURVES OF CONSTANT WIDTH

Tool(s) Geometers Sketchpad   Data File(s)   gsp13   (Mac & PC)

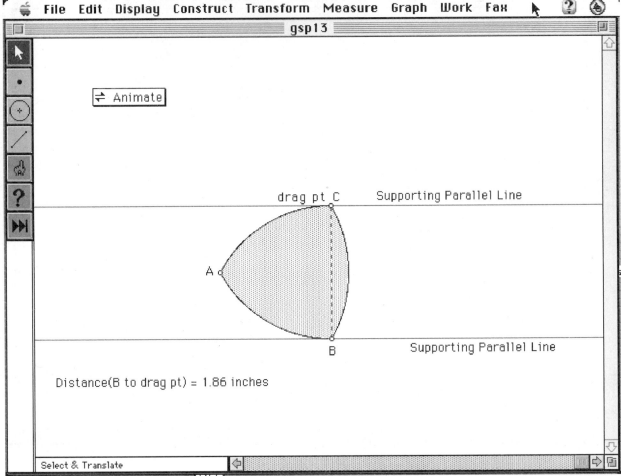

Focus

A curve of constant width is a convex oval in which the distance between parallel tangents is constant.

Tasks

1. Double click the Animate button. A single click stops the animation. The curve may be resized by moving points A and B. Discuss your observations.

2. Imagine that the parallel lines were fixed and the curve turned between them. What shape would the turning curve trace out between the lines? If you had a drill that would behave in this manner, what shape holes would it drill?

ORTHOGONAL CIRCLES

First Case: One Point on the Circle and One Point in the Interior of the Circle
Tool(s) Geometers Sketchpad   Data File(s)   gsp14   (Mac & PC)

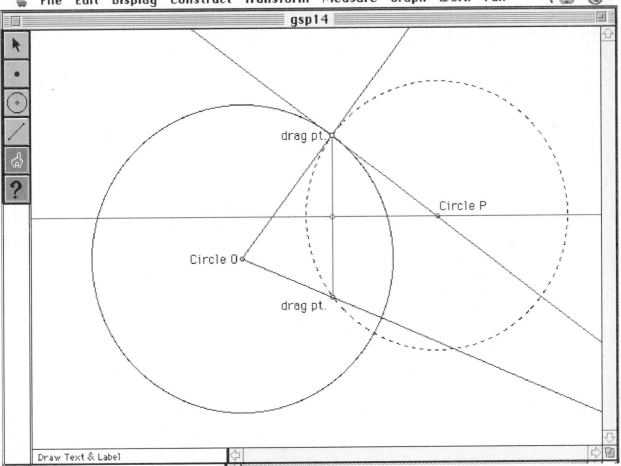

Focus

Circle P is constructed orthogonal to Circle O, given one point on Circle O and a point in the interior of Circle O.

Tasks

1. Construct a second set of tangents to both circles at their other point of intersection. What angles are formed? When two circles are found to be orthogonal at one point of intersection, will they always be orthogonal at the other point of intersection? Why or why not?

2. Describe what happens when you move the interior drag pt outside of Circle O. Is this construction valid whether the given point is in the interior or exterior of the circle? Why or why not?

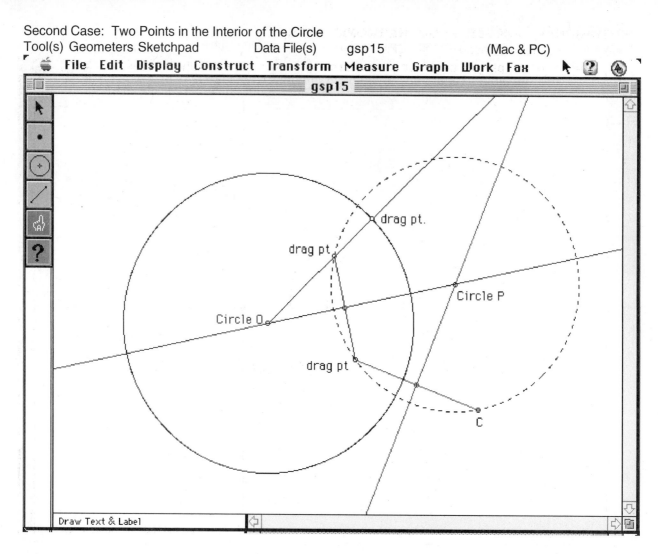

Focus

Circle P is constructed orthogonal to Circle O, given two points in the interior of Circle O.

Tasks

1. What happens to Circle P when the given points move closer to Circle O? To the center of Circle O?

2. Describe what happens when you move the given points outside of Circle O. Does the construction appear to be valid whether the given points are in the interior or exterior of the Circle O?

Tool(s) Geometers Sketchpad   Data File(s)   gsp16   (Mac & PC)

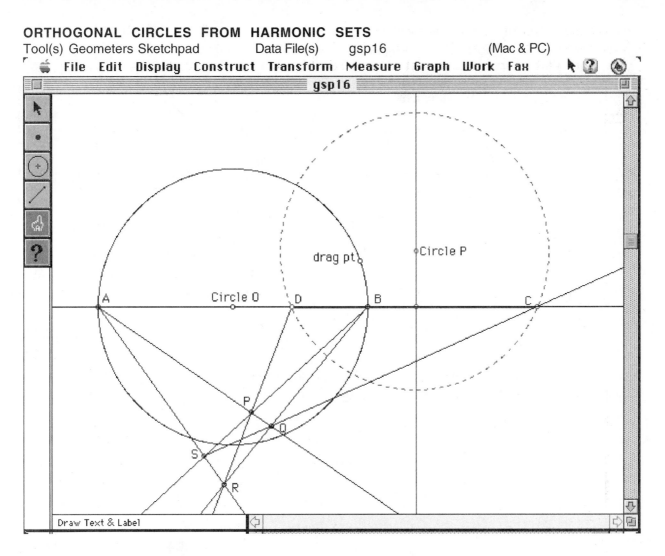

Focus

Point D is given as one of the points in the interior of Circle O through which orthogonal Circle P must pass. AB, a diameter of Circle O, is drawn containing point D. The task is to find a point C on line AB that is also on Circle P.

Cross Ratio: In his book *Mathematical Collection,* Pappus of Alexandria (300 A.D.) proved that if four concurrent rays are cut by two transversals (Figure 1.1), their cross ratios are equal.

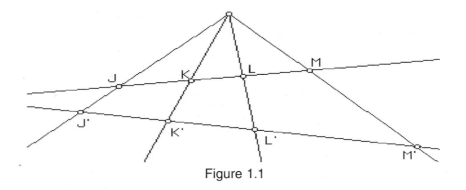

Figure 1.1

In the context of Figure 1.1, the cross ratios involved are computed as (JK/JM) / (−LK/LM) and (J´K´/J´M´)/(−L´K´/L´M´). Pappus' discovery has far reaching consequences in projective geometry where it is restated as a fundamental theorem, "The cross ratio of four collinear points is invariant under projective transformations." For now, it is sufficient to note that the ratios may be computed directly from the measurements.

Harmonic Sets: The simplest and most familiar arithmetic progression is 1, 2, 3, 4, 5, ... Successive elements in arithmetic progressions are separated from one another by a constant difference, in this case 1. Elements in harmonic progression are reciprocals of elements in arithmetic progression. For instance, the first harmonic progression encountered by most students of mathematics is 1/1, 1/2, 1/3, 1/4, 1/5 ... , and so on. If four consecutive elements of a harmonic progression are used to define the endpoints of segments like those in Figure 1.1, the cross ratio of the points may be computed. For example, the elements {1/4, 1/3, 1/2, 1} may be used as coordinates along the segment shown in Figure 1.2. The cross ratio of this and all harmonic sets is the same, −1.

1/4   1/3   1/2   1

Figure 1.2

Complete Quadrangle: A complete quadrangle is a set of four points (P, Q, R, S), no three collinear, and the six lines determined by the four points (Figure 1.3).

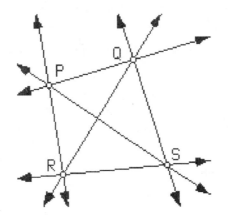

Figure 1.3

Four points, A, B, C, and D, form a harmonic set H (AB, CD) if there is a complete quadrangle in which two opposite sides pass through A, two other opposite sides pass through B, and the remaining two sides pass through C and D, respectively. In gsp16, the six lines defined by quadrangle PQRS intersect the diameter of Circle O at points A, B, C, and D, guaranteeing that these points form a harmonic set. Adjusting the location of point D on this diameter results in a new position for point C as determined by quadrangle PQRS.

Orthogonal Circles: If two intersecting circles divide a line segment containing the center of one of the circles harmonically (cross ratio = −1), then the circles are orthogonal.

Tasks

1. Verify that the circles are orthogonal by constructing tangents at the points of intersection.

2. Move point D along line AB. State a conjecture concerning the movement of point C. To prove your conjecture false, what sort of counter example would be necessary?

3. Using the arrow icon, resize Circle O using the drag pt. Discuss your observations.

4. Using the arrow icon, slide the center of Circle P along the secant containing its diameter. Discuss your observations.

5. Why is the movement of the center of Circle P constrained? Discuss why this procedure generates many orthogonal circles instead of just one.

6. Move point D outside of Circle O. Discuss your observations.

7. State a conjecture concerning the possible locations of the center of Circle P. To prove your conjecture false, what sort of counter example would be necessary?

HARMONIC SETS AND THE CROSS RATIO

Tool(s) Geometers Sketchpad   Data File(s)   gsp17   (Mac & PC)

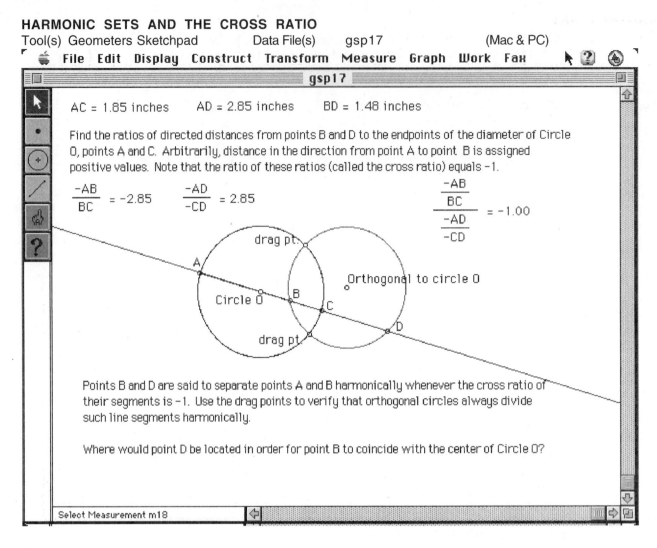

Within the sketch:

AC = 1.85 inches   AD = 2.85 inches   BD = 1.48 inches

Find the ratios of directed distances from points B and D to the endpoints of the diameter of Circle O, points A and C. Arbitrarily, distance in the direction from point A to point B is assigned positive values. Note that the ratio of these ratios (called the cross ratio) equals −1.

$$\frac{-AB}{BC} = -2.85 \qquad \frac{-AD}{-CD} = 2.85 \qquad \frac{\frac{-AB}{BC}}{\frac{-AD}{-CD}} = -1.00$$

drag pt

Orthogonal to circle O

A

Circle O   B   C

D

drag pt

Points B and D are said to separate points A and B harmonically whenever the cross ratio of their segments is −1. Use the drag points to verify that orthogonal circles always divide such line segments harmonically.

Where would point D be located in order for point B to coincide with the center of Circle O?

Select Measurement m18

Focus

Two orthogonal circles divide the interior portions of a secant passing through the center of one of the circles harmonically.

Tasks

1. Using the arrow icon, move the drag pts and observe the effect on the ratios of distances displayed to the left of the circle. State a conjecture based on your observations. To prove your conjecture false, what sort of counter example would be necessary?

2. Using the arrow icon, move the drag pts and observe the effect on the cross ratio displayed to the right of the circle. State a conjecture based on your observations. To prove your conjecture false, what sort of counter example would be necessary?

26

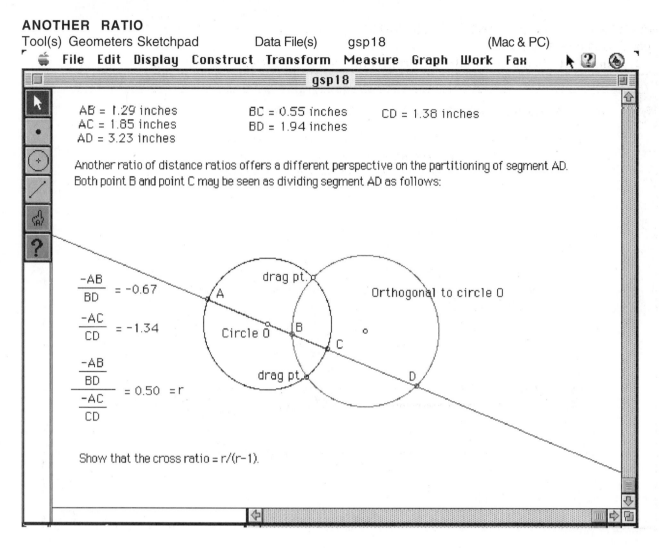

File   Edit   Display   Construct   Transform   Measure   Graph   Work   Fax

Focus

This activity provides a different analysis of the manner by which two orthogonal circles divide the interior portions of a secant passing through the center of one of the circles.

Tasks

1. Using the arrow icon, move the drag pts and observe the effect on the ratios of distances displayed. State a conjecture based on your observations. To prove your conjecture false, what sort of counter example would be necessary?

2. Show that the cross ratio is equal to $r / (r - 1)$.

CHAPTER 2

EXPLORING OTHER GEOMETRIES

From the time of Euclid (300 B.C.) until the 19th century, the study of geometry rested on the bedrock of Euclid's *Elements*. The foundation on which the *Elements* rested was Euclid's five postulates. Expressed in simple terms, they are ...

1. Between any two points there is a unique straight line.
2. Line segments of any length may be constructed on any line.
3. Circles of any center and radius may be constructed.
4. All right angles are congruent.
5. Through a given point not on a given line, exactly one parallel may be drawn (Playfair's version).

Beginning with these postulates, some terminology, and a few common notions, Euclid constructed a system of geometrical thought that has endured for centuries and served as a basis for teaching axiomatic thinking to a hundred generations of students. Perhaps because Euclid constructed his first 28 propositions (theorems) without resorting to the fifth postulate, and perhaps because it is less self-evident than the others, scholars through the centuries have attempted to prove that Euclid's 5th postulate could be derived from the first four, making it a theorem. The most famous attempt to do so was by the Italian Jesuit priest Girolamo Saccheri (1667 – 1733).

Saccheri began by constructing a quadrilateral ABCD in which angles A and B are right angles and sides AD and BC are congruent. Using congruence theorems found in the first 28 of Euclid's propositions, Saccheri proved that angles C and D are congruent. He then stated that angles C and D are jointly, either acute, right, or obtuse. His objective was to rule out the first and last options using logic, leaving the middle option as the only possible choice. This result, would, in effect, prove the parallel postulate.

Figure 2.1

Saccheri succeeded in his attempt to show that the obtuse option was impossible. When he attempted the acute option, however, he could only produce a lame argument. Nevertheless, he believed that he had achieved his goal. Had he only faced facts and conjectured that a consistent axiomatic system could be generated from the first four of Euclid's postulates and an acute angle postulate, he would have been credited as the discoverer on non-Euclidean geometry.

The first mathematicians to suspect that consistent alternatives to Euclidean geometry existed were Carl Friedrich Gauss (1777 – 1855) of Germany, Nikolai Lobachewsky (1793 – 1856) of Russia, and Janos Bolyai (1802 – 1860) of Hungary. Working independently, these men discovered geometries in which the 5th postulate was fundamentally different from Euclid's. The term non-Euclidean is used to designate geometries based on these differences. In each case, the 5th postulate may be stated as beginning "Through a given point not on a given line, ... " and finishing ...

Hyperbolic 5th Postulate:	"... infinitely many parallels may be drawn."
Euclidean 5th Postulate:	"... exactly one parallel may be drawn."
Elliptic 5th Postulate:	"... no parallels may be drawn."

Although Gauss, Lobachewsky, and Bolyai first investigated the validity of non-Euclidean axiomatic systems, it was Henri Poincare of France (1854 – 1912) who created the most popular model for exploring hyperbolic geometry, the Poincare disk. To understand Poincare's model, consider the lipstick case shown in Figure 2.2. Seen from the side, the lipstick itself is encased in a cylindrical tube with a circular end. Imagine that two-dimensional inhabitants move about the surface of the lipstick, much as germs might. These inhabitants have no concept of up or down, just sideways movement across the surface of the lipstick. To such creatures, space is not curved like the lipstick, but flat like the circular end of the case. An individual inhabitant moving from point A to B would think of his motion in terms of a line from A' to B'. If that distance is assumed to be one unit of length, and if the lipstick and cylindrical case are asymptotic to each other, the inhabitant might continue his journey along the same line, going past B the same distance to a point C. To the inhabitant of this space, $A'B' = B'C'$. From our perspective (looking down on the circular end of the lipstick case, not from the side as in Figure 2.2), the distance $B'C'$ would appear shorter than $A'B'$. Furthermore, any object moving from the center toward the circle would appear to shrink as it receded deeper into the lipstick tube. From our perspective, the inhabitants of hyperbolic space live in a finite, bounded universe. From their perspective, the universe is infinite. The tools NonEuclid and Hyperbolic MacDraw enable you to explore hyperbolic space as modeled in the Poincare disk.

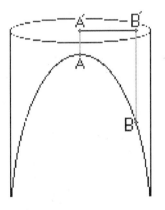

Figure 2.2

HYPERBOLIC GEOMETRY

INTRODUCTION TO NONEUCLID

Tool(s) NonEuclid 4.0b   Data File(s)   none   (Mac & PC)

File   Edit   Options   Constructions   Measurements   Help

```
Plotted Point A
Plotted Point B
Length AB = 3.723
Plotted Point C
Length BC = 4.616
Length CA = 5.531
Triangle ABC
    Length AB = 3.723
    Length AC = 5.531
    Length BC = 4.616
    Angle BAC = 11°
    Angle ABC = 28°
    Angle ACB = 4°
    Angle Sum =  44°
    Area =  136°
```

Focus

NonEuclid includes a set of introductory activities that highlight the basic features of hyperbolic geometry as demonstrated on the Poincare disk.

Tasks

The following activities are available under the Help pull-down menu.

Introduction	My First Triangle	Menu Commands
Distance	Parallel Lines	What To Do
Angles	Triangle	Isosceles Triangle
Equilateral Triangle	Right Triangle	Parallelogram
Rhombus	Rectangle	Square
Circle	Euclid's Postulates	

As you work on each activity, look for answers to the following questions.

1. In any geometry, the term 'line' is reserved for the shortest curve between two points. What is the apparent angular relationship between lines in the Poincare disk and the boundary of the disk?

2. If you lived in a hyperbolic universe, would you believe it to be finite or infinite? Design an experiment that would tell you whether your space was "flat" or "curved."

3. If "parallel" lines are defined as non – intersecting, how many parallels may be drawn to a given line through a point not on the given line?

4. Investigate possible angle sums for hyperbolic triangles. What is the minimum possible angle sum? Under what conditions do such triangles occur? Is there a maximum possible angle sum?

5. Construct several right triangles and measure their sides. Using a calculator, determine whether the Pythagorean Theorem holds in hyperbolic geometry.

6. Is it possible to construct a rectangle in hyperbolic geometry? Why or why not?

7. How is area measured in hyperbolic geometry? Why are conventional units of area impossible in hyperbolic space?

8. Construct several unit circles within the Poincare disk. How does the appearance of circles drawn near the boundary of the disk differ from circles drawn near the center of the disk? If you could move a ruler from the center of the disk toward the boundary, what changes would you observe in the ruler?

31

9. How does hyperbolic geometry differ from Euclidean geometry? Using NonEuclid, construct hyperbolic counter examples to the following theorems in Euclidean geometry. Sketch the results in the space provided.

- If two lines are parallel to a third line, they are parallel to each other.

- If two parallel lines are cut by a transversal, the alternate interior angles are congruent and the corresponding angles are congruent.

- The measure of an exterior angle of a triangle is equal to the sum of the measures of the remote interior angles.

- Any angle inscribed in a semicircle is a right angle.

- The altitude to the hypotenuse of a right triangle forms two triangles, each of which is similar to the original triangle and to each other.

10. How is hyperbolic geometry similar to Euclidean geometry? Using NonEuclid, illustrate the following theorems common to both geometries. Sketch the results in the space provided.

- You can construct an equilateral triangle.

- The base angles of an isosceles triangle are congruent.

- Vertical angles are congruent.

- Angle-Side-Angle establishes congruence between triangles.

- Angle-Angle-Side establishes congruence between triangles.

Tool(s) Hyperbolic MacDraw   Data File(s)   none   (Mac)

File   Edit   Model   Tool   Pen   Fill   Fox   1:31 PM

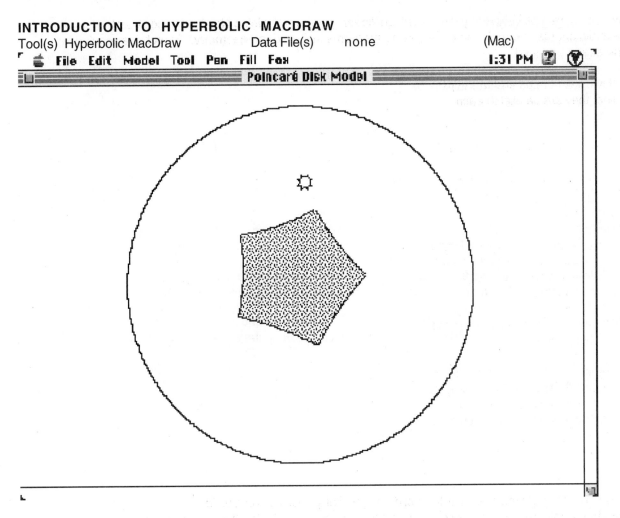

Poincaré Disk Model

Focus

Hyperbolic MacDraw offers a different set of tools than NonEuclid for the investigation of hyperbolic space.

Tasks

1. The regular pentagon shown above was created using the polygon tool ⚙ in the Tools pull-down menu. To create your own version of the figure, select the polygon tool, drag the ⚙ into the Poincare disk and click the mouse, using the ⚙ icon as a stamp to position the polygon. Enter 5 for the Number of Sides in the Polygon Specification dialog box and click OK. Repeat this procedure, positioning several other regular pentagons within the Poincare disk. Describe and explain the result.

2. Using the Select All option under the Edit pull-down menu, erase all of your pentagons using the Backspace key. Create a regular pentagon at the center of the Poincare disk as before, then select the hand icon ⟨ᵐᵐⁿ⟩ from the Tools pull-down menu. Position the hand icon ⟨ᵐᵐⁿ⟩ over the pentagon and move the mouse around with the button down. Describe and explain the result.

3. Using the Select All option under the Edit pull-down menu, erase all of your pentagons using the Backspace key. Select the freehand polygon tool located in the lower left-hand corner of the Tools pull-down menu. Drag the freehand polygon icon onto the Poincare disk, where it will look like ⟨⟩. Draw a closed figure. When you return to the place where you began to draw, the figure will close and fill with whichever pattern is selected in the Fill pull-down menu. Select the rotation tool ↺ in the Tools pull-down menu and position it over the figure. Describe and explain what happens when you depress the mouse button. Now position the rotation tool ↺ outside the figure and depress the mouse button. Describe and explain what happens.

4. What features of hyperbolic space does Hyperbolic MacDraw offer that you wished were available in NonEuclid?

5. What features of NonEuclid do you wish were available in Hyperbolic MacDraw?

34

TILING THE HYPERBOLIC PLANE

Tool(s) NonEuclid 4.0b Data File(s)   Web of Congruence   (Mac & PC)

File   Edit   Options   Constructions   Measurements   Help

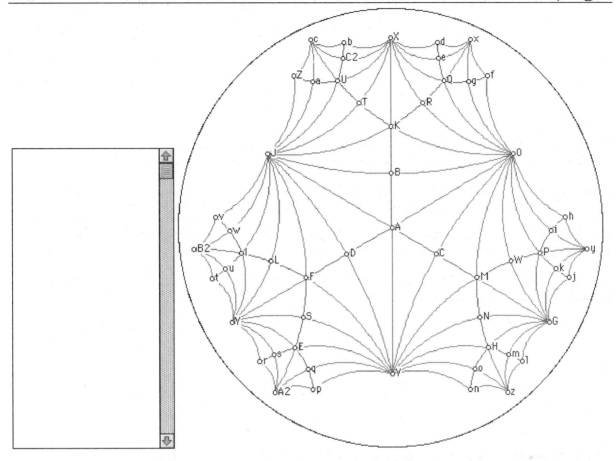

Focus

A figure is said to "tile" the plane when it covers the plane without overlapping itself or leaving gaps. This activity takes advantage of the Web of Congruence example that comes with NonEuclid 4.0b.

Tasks

1. Open the file Web of Congruence.   Using the Measure Triangle option under the Measurements, measure several triangles. Are the corresponding segments congruent? Are the corresponding angles congruent? Do the triangles have the same angle sum? The same area?

2. Find a different triangle that may be used to tile the plane. Demonstrate that your triangle does tile 2-dimensional hyperbolic space.

35

FINITE GEOMETRY

Finite geometries consist of a finite number of points and their relationships. Finite geometries provide a manageable context in which to investigate the nature of axiomatic systems. In general, finite geometries are explored using models that embody all of the attributes of the axiomatic system. For instance, a three point geometry might be represented using three labeled points on paper, three letters arranged as rows or columns, and so on (Figure 2.3).

Model #1
    The points are the dots.
    The lines are pairs of points
      connected by dashes.

Model #2:
    The points are the letters A, B, C.
    The lines are pairs of letters
      arranged in columns.

    A B C
    B C A

Figure 2.3

For most people, study of the logical features of an axiomatic system is greatly facilitated by models. The principal problem associated with the use of models is that one must avoid importing inappropriate ideas and symbols from Euclidean geometry. For instance, lines and line segments do not exist in the Euclidean sense in any finite geometry. This fact rules out the existence of angles, polygons, and other familiar Euclidean objects and their measurements (length, area, and so on). Lines in finite geometries consist of finite sets of points understood to stand in a particular relationship. While these "lines" may be sketched as Euclidean lines, it is understood that the only points under consideration are those belonging to the finite geometry.

Among the many features of axiomatic systems, the most fundamental questions are ...

1. Is the system consistent? If no contradictions can be deduced from the axioms, a system is consistent.
2. Is the system independent? If no axiom is implied by the other axioms, the set of axioms is independent.
3. Is the system complete? If no additional axioms can be added to the system, or equivalently, if every properly posed question asked of the system may be answered, the system is complete.
4. Are two axiomatic systems with different sets of axioms equivalent? If both systems lead to the development of the same body of knowledge, they are equivalent.

It is the very sparseness of finite geometries that make them such appealing models for the study of axiomatic systems. Another benefit is their unfamiliarity. Too often, when studying the logical features of some more familiar set of objects, our familiarity limits our insight and dulls our thinking. Finite geometries provide "tiny" unexplored universes in which to investigate the nature of thinking itself.

THREE POINT GEOMETRY

Tool(s) Geometers Sketchpad   Data File(s)   gsp19   (Mac & PC)

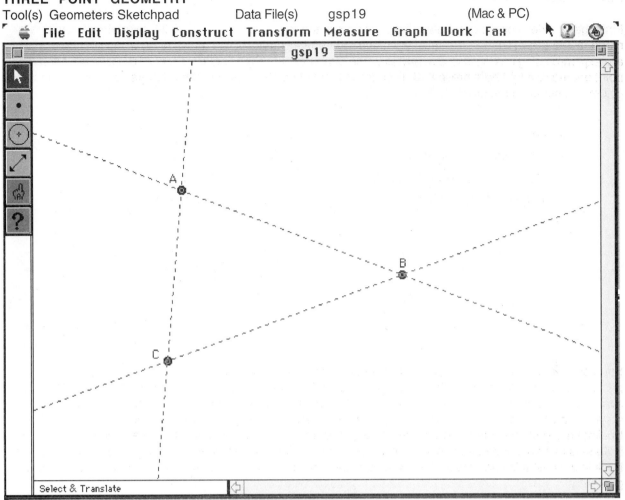

Focus

The postulates for three point geometry may be stated as ...
1. There exist exactly three points.
2. Two distinct points are on exactly one line.
3. Not all the points are on the same line.
4. Two distinct lines are on at least one point.

Tasks

1. Verify that the finite geometry represented in the figure satisfies the axioms for three point geometry.

2. State at least one additional conjecture that might be proven as a theorem in this geometry. To prove your conjecture false, what sort of counter example would be necessary?

FOUR POINT GEOMETRY

Focus

The postulates for four point geometry may be stated as ...

1. There are exactly four points.
2. Any two distinct points are on exactly one line.
3. Each line is on exactly two points.

Tasks

1. Create a model of four point geometry using the GSP.

2. State at least two conjectures about four point geometry that might be proven as theorems. To prove each conjecture false, what sort of counter example would be necessary?

FINITE GEOMETRY OF PAPPUS

Tool(s) Geometers Sketchpad   Data File(s)   gsp20   (Mac & PC)

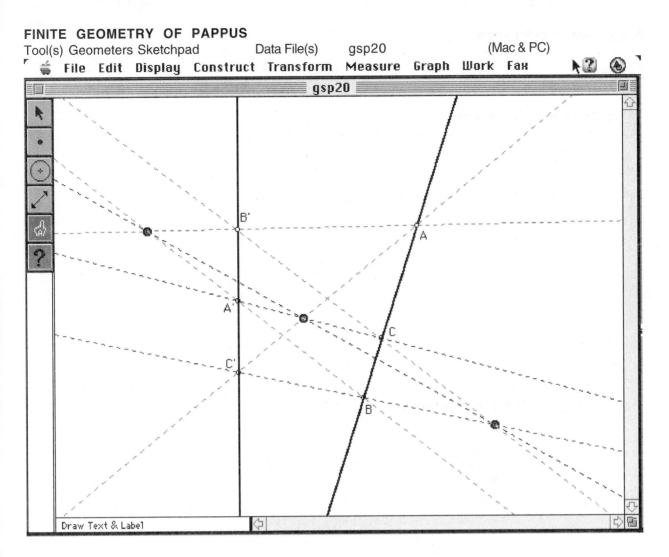

Focus

The finite geometry of Pappus (ca. 300) is based on nine points constructed using a theorem from Euclidean geometry called the Theorem of Pappus: If A, B, and C are three points on one line, and if A', B', C' are three points on a second line, then the points $D = AC' * CA'$, $E = AB' * BA'$, and $F = BC' * CB'$ are collinear.

Tasks

1. Using the GSP's arrow icon, drag points A, B, C, A', B', and C' around. Does the theorem hold if you change the order of points A, B, and C on the line?

2. How many labeled points are there on each line? State a conjecture based on your observations. To prove your conjecture false, what sort of counter example would be necessary?

3. How many lines are concurrent at each labeled point? State a conjecture based on your observations. To prove your conjecture false, what sort of counter example would be necessary?

4. Given a line and a point not on the line, how many other lines contain the given point and given line? State a conjecture based on your observations. To prove your conjecture false, what sort of counter example would be necessary?

5. Given a line and a point not on the line, how many lines concurrent with the given point do not contain labeled points on the given line? How is this situation similar to the notion of parallel lines in Euclidean geometry? State a conjecture based on your observations. To prove your conjecture false, what sort of counter example would be necessary?

FINITE GEOMETRY OF DESARGUES

Tool(s) Geometers Sketchpad   Data File(s)   gsp21   (Mac & PC)

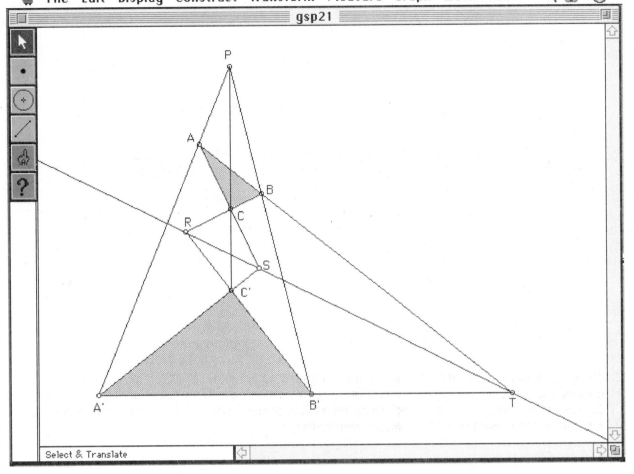

Focus

The finite geometry of Gerard Desargues (1591 – 1662) involves a total of ten points on ten lines.

Tasks

1. How many labeled points are there on each line? How many lines are concurrent at each labeled point?

2. What relationship exists between corresponding vertices and point P?

3. What relationship exists between corresponding sides and line RST?

CHAPTER 3

TRANSFORMATION GEOMETRY

Geometric transformation was first studied systematically in ancient Greece, where repetitive patterns called friezes were used in the decoration of buildings. Today, geometric transformations play an essential role in the work of mechanical engineers, architects, and computer scientists. From entertainment to scientific visualizations, computer graphics now play a critical role in communicating ideas and motivating viewer response. Virtually all such applications involve some use of geometric transformations.

ISOMETRIES

Most students begin their study of linear transformations with an investigation of basic isometries: translations (slides), reflections (flips), rotations (turns), and glide reflections. Each of these transformations preserves collinearity of points, concurrence of lines, segment length, and angle measure. As a result, an object and its image under an isometry are congruent. A study of isometries should reveal much more than this, however. Isometries may be conceived as operating on and transforming the entire Euclidean plane. When approached in this manner, the study of isometries and linear transformations in general may be enlarged to include such topics as invariant points and lines and the matrix representation of linear transformations.

The English mathematician Arthur Cayley (1821 – 1895) developed matrices and matrix algebra in the context of describing linear transformations. The first benefit of Cayley's new notation was to simplify the representation of linear transformations. For example, a general linear transformation written in algebraic form involves two equations:

$$x' = ax + by + e$$
$$y' = cx + dy + f$$

Using matrix notation, both of these equations are combined into a single matrix equation:

$$\begin{bmatrix} a & b & e \\ c & d & f \\ 0 & 0 & 1 \end{bmatrix} \cdot \begin{bmatrix} x \\ y \\ 1 \end{bmatrix} = \begin{bmatrix} x' \\ y' \\ 1 \end{bmatrix}$$

Since Cayley, matrix algebra has become a powerful tool in geometry and applied mathematics. The tools used in this chapter demonstrate the elegance and power of matrix algebra in the study of geometric transformations.

TRANSLATIONS (SLIDES)

Tool(s) Geometers Sketchpad   Data File(s)   gsp22   (Mac & PC)

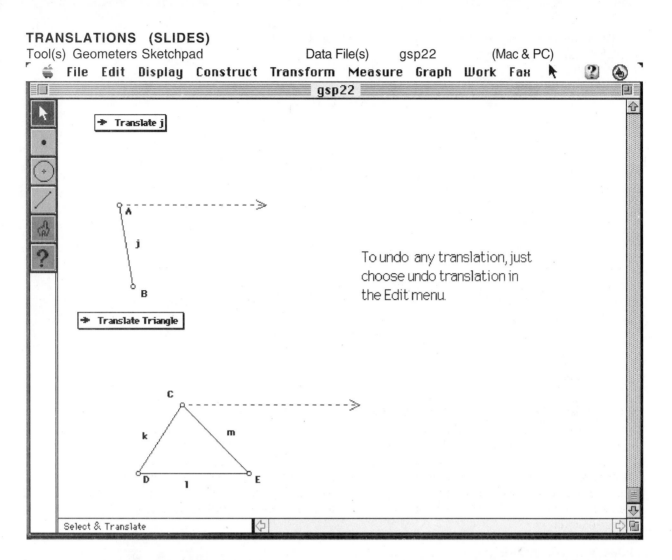

Focus

A translation (slide) moves an object a given distance in a given direction, often indicated by a slide arrow.

Tasks

1. Double click on the Translate j action key. Describe the action.

2. Measure the length and slope of j before and after the translation.

3. Measure AA´ and B B´.

43

4. Sketch the segment *j*, the slide arrow, and an additional line *l* that would be invariant (slide along itself) under the given translation.

5. Double click on the Translate Triangle action key. Describe the action.

6. Measure the sides and angles of the triangle before and after the translation.

7. Measure the area of the triangle before and after the translation.

8. Measure CC´ and EE´.

9. Sketch the triangle, the slide arrow, and an additional line *l* that would be invariant under the given translation.

10. What action would reverse a translation? Or, what is the inverse of a translation?

REFLECTIONS (FLIPS)

Tool(s) Geometers Sketchpad   Data File(s)   gsp23   (Mac & PC)

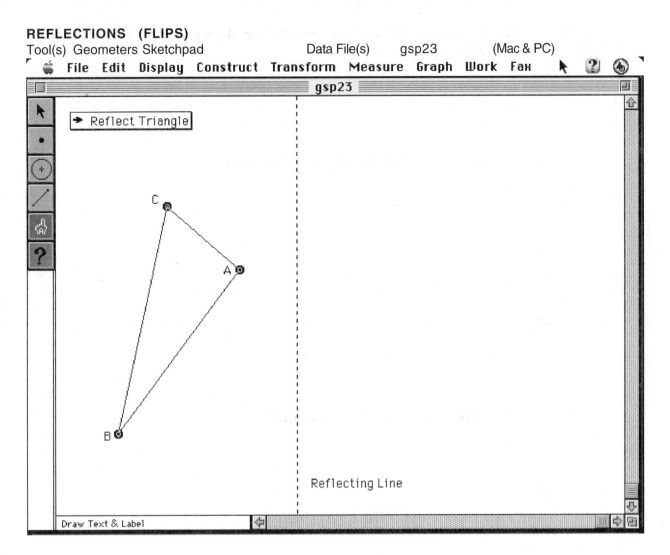

Focus

A reflection produces a mirror image of an object on the opposite side of a reflecting line.

Tasks

1. Double click on Reflect Triangle. Double click again. What is the inverse of a reflection?

2. Measure and compare the distances from each point of the original triangle and reflected triangle to the reflecting line.

3. Sketch the original triangle, reflecting line, and at least one additional line that is invariant under this reflection.

ROTATIONS (TURNS)

Tool(s) Geometers Sketchpad   Data File(s)   gsp24   (Mac & PC)

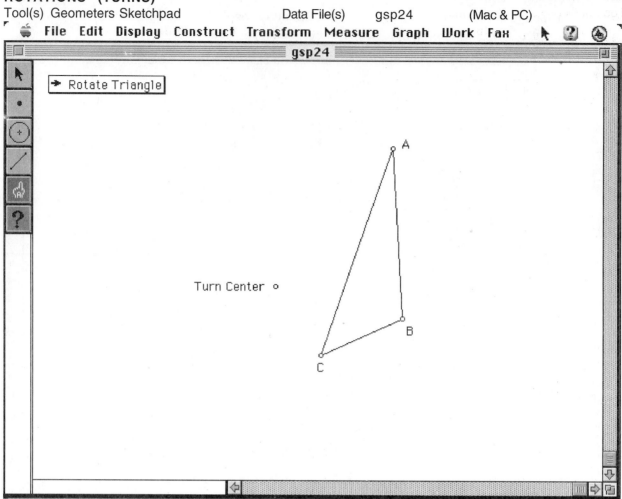

Focus

A rotation is defined by a turn center and a turn angle.

Tasks

1. Double click on the Rotate Triangle action button.   Double click again. What is the inverse of a rotation?

2. Describe a method for determining the turn angle of a rotation given an object and its image.

3. Identify any invariant points or lines associated with this rotation.

46

GLIDE REFLECTIONS

Tool(s) Geometers Sketchpad   Data File(s)   gsp25   (Mac & PC)

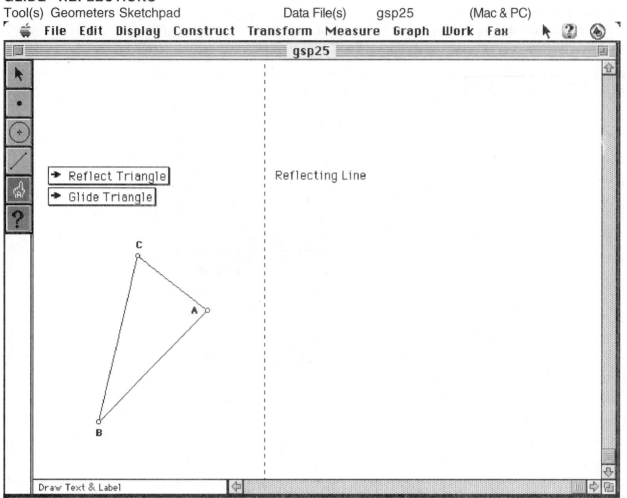

Focus

Glide reflections involve both translation and reflection.

Tasks

1. Double click on the Reflect Triangle action key then the Glide Triangle action key.   Double click on the Glide Triangle action key, then the Reflect Triangle action key. Compare the final results obtained using both sequences of actions.

2. Identify any points or lines that are invariant under this glide reflection.

LINEAR TRANSFORMATIONS

In addition to translating, reflecting, and rotating the plane, linear transformations may strain, shear, or dilate. These transformations are illustrated below using the software tool MoVil.

- The first view (Figure 3.1) is of a unit square and the x − y coordinate system.

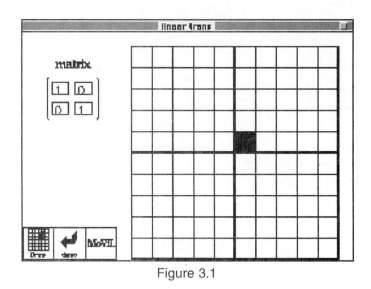

Figure 3.1

- In Figure 3.2, the unit square has been transformed into a 2 x 3 rectangle. The 2 x 2 matrix for this transformation appears to the left of the rectangle and involves a strain (stretch or compress) in the x − direction by a factor of two and a strain in the y − direction by a factor of three.

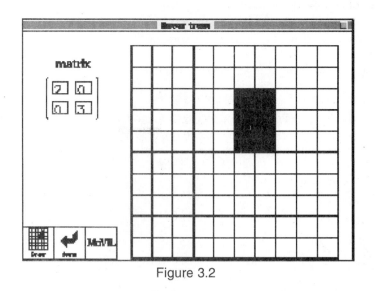

Figure 3.2

The scaling factors are apparent in the matrix algebra for this transformation:

$$\begin{bmatrix} 2 & 0 \\ 0 & 3 \end{bmatrix} \cdot \begin{bmatrix} x \\ y \end{bmatrix} = \begin{bmatrix} 2x \\ 3y \end{bmatrix}$$

48

- A shear shifts points parallel to a shear axis by an amount proportional to the points' distances from the shear axis. In this case, the shear axis is the x – axis.

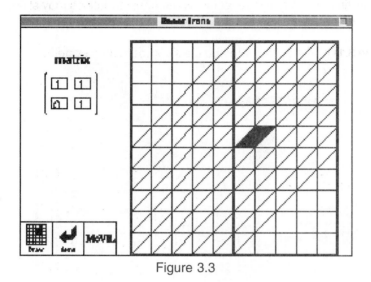

Figure 3.3

The shearing effect seen in Figure 3.3 is generated by the element in the first row, second column of the transformation matrix:

$$\begin{bmatrix} 1 & 1 \\ 0 & 1 \end{bmatrix} \cdot \begin{bmatrix} x \\ y \end{bmatrix} = \begin{bmatrix} 1x + 1y \\ 1y \end{bmatrix}$$

- A dilation expands or contracts, much as a photographic expansion or reduction. In this case the dilation is by a factor of two about the origin.

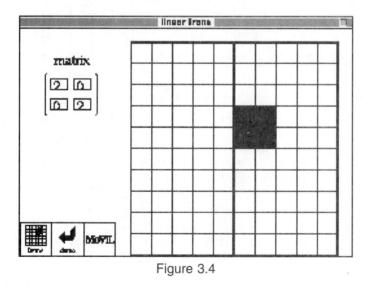

Figure 3.4

The matrix algebra for the dilation seen in Figure 3.4 is:

$$\begin{bmatrix} 2 & 0 \\ 0 & 2 \end{bmatrix} \cdot \begin{bmatrix} x \\ y \end{bmatrix} = \begin{bmatrix} 2x \\ 2y \end{bmatrix}$$

In general, linear transformations may be viewed as combinations of these basis transformations: translations, reflections, rotations, strains, shears, and dilations. Two examples follow:

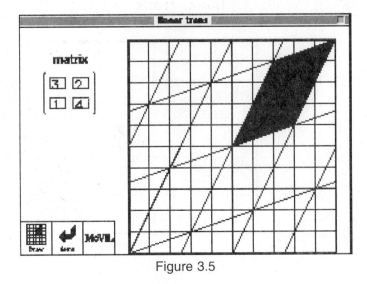
Figure 3.5

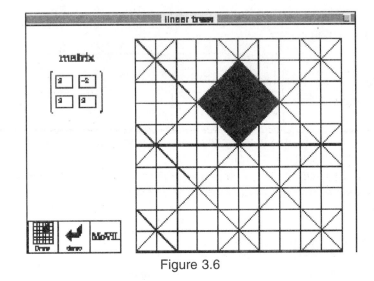
Figure 3.6

MoVil creates the impression that linear transformations occur as continuous motions from an initial state to a final state. In fact, linear transformations do not involve "motion" in this sense. Each linear transformation relates an initial and a final state but does not imply intermediate states. MoVil adds intermediate states and animates the transitions between these states in order to present mathematical concepts in a context familiar to most students, motions in the physical world. In this context, students may more easily consider questions of the sort "Are there any fixed points or fixed lines?" For instance, when viewing a rotation with MoVil, most students readily observe that the only fixed point is the turn center. Or, when viewing a reflection, it is apparent to most students that the reflecting line is pointwise invariant and that all lines perpendicular to it are invariant.

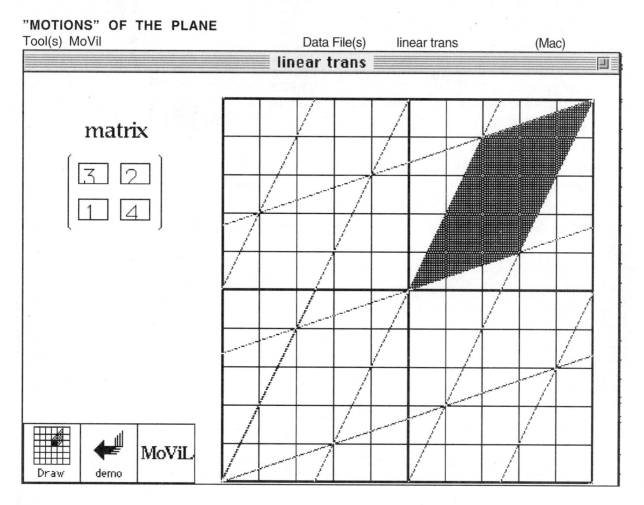

Focus

In general, linear transformations determined by non-singular, 2 x 2 matrices with integers for elements map squares onto parallelograms.

Tasks

1. Run the "linear trans" demo that comes with MoVil.   Write the matrix of a dilation.

2. Write a matrix for a reflection in the y-axis. Discuss the existence of fixed points under this transformation. What should you look for in the MoVil animation to support your conjecture? Prove your conjecture analytically.

3. Write a matrix that produces a 180 degree rotation about the origin. Discuss the existence of fixed points under this transformation. What should you look for in the MoVil animation to support your conjecture? Prove your conjecture analytically.

FRIEZE PATTERNS

Friezes are designs that repeat a geometric pattern along a single dimension. In the figure below, rectangular boxes are laid out in a row with a center line passing through each. The midpoint of the portion of the center line in the first box is marked with a dot. Any pattern entered in the first box is translated to the right as many times as necessary to produce a frieze of a particular length.

Frieze Grid

Tool(s) Geometers Sketchpad   Data File(s)   gsp26   (Mac & PC)

Focus

A study of the mathematics of frieze patterns focuses on the symmetries used in creating the pattern in the first box.

Tasks

1. Create a separate frieze based on each of the patterns shown in Figure 3.7. Sketch the pattern into the first box using the segment tool. After marking the width of the box as a vector, translate the pattern into the second box. Save each frieze as a separate Sketchpad file.

LLLLLLL
NNNNNN
DDDDDD
UUUUUU
HHHHHH
VAVAVA
LᒋLᒋLᒋL

Figure 3.7

- LLLLLLLL

- NNNNNN

- DDDDDD

- UUUUUU

- HHHHHH

- VAVAVA

- LᒋLᒋLᒋL

Tool(s) Geometers Sketchpad   Data File(s)   gsp27   (Mac & PC)

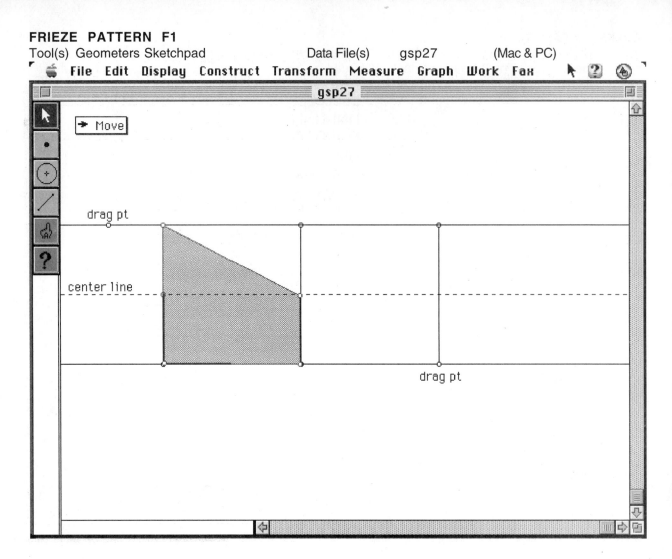

Focus

Translating a pattern from box to box.

Tasks

1. Double click on the Move action button.   Measure and compare the pattern before and after the move. Discuss your findings.

2. Describe the pattern and identify any points or lines of symmetry.

3. Which pattern in the Frieze Grid activity has the same symmetries?

FRIEZE PATTERN F2

Tool(s) Geometers Sketchpad   Data File(s)   gsp28, gsp29   (Mac & PC)

Focus

F2 frieze patterns have a single point of symmetry and no lines of symmetry.

Tasks

1. Create a quadrilateral like the one shown in the first row of gsp28 and fill its interior as shown in the second row.

2. Mark the midpoint of the center line as the turn center for a rotation.

3. Select the vertices of the quadrilateral and its interior, then rotate 180 degrees as shown in row three above.

4. Manipulate the vertices so that the original figure and its rotation match perfectly and so that there are no lines of symmetry.

5. Compare your figure to the letter N found in Figure 3.8 (gsp29). Do they have the same symmetries?

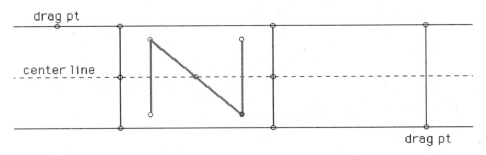

Figure 3.8

OTHER FRIEZE PATTERNS

Tool(s) Geometers Sketchpad   Data File(s)   gsp26   (Mac & PC)

Focus

Each frieze pattern has a different set of symmetries.

Tasks

1. Pattern F^1_1 has the center line as a line of symmetry but no points of symmetry, like the pattern DDDDDD. Construct a frieze based on this pattern.

2. Pattern F^2_1 has a line of symmetry perpendicular to the centerline at the midpoint, like the pattern VVVVVV. Construct a frieze based on this pattern.

3. Pattern F^1_2 is symmetric about the center line and has a point of symmetry at the midpoint, like the pattern HHHHHH. Construct a frieze based on this pattern.

4. Pattern F^2_2 is symmetric about a line perpendicular to the centerline 1/4 of the width of the box from the left edge and a point of symmetry at the midpoint, like the pattern VΛVΛVΛ. Construct a frieze based on this pattern.

5. Pattern F^3_1 is a glide reflection. Construct a frieze based on this pattern.

6. Examine stationary samples for borders that are frieze patterns. Classify each into one of the seven frieze patterns.

TESSELLATIONS

A tessellation is a geometric pattern that covers the plane without gaps or overlaps. Familiar objects that can be used in this manner are squares, rectangles, equilateral triangles, and regular hexagons. Figure 3.9 illustrate the concept.

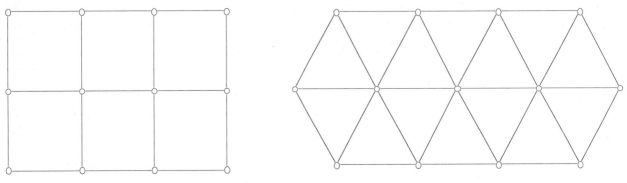

Figure 3.9

Creating tessellations that remind us of objects, animals, and people is as much an art as a science. Figure 3.10 suggests a pinwheel or propeller. Based on tessellating equilateral triangles, side AB is redefined by drawing curve A-A1-A2-A3-M, where M is the midpoint of AB, and performing a glide reflection of that curve from M to B. When this transformation is repeated on the other two sides of the triangle, a three-bladed, propeller-like shape emerges. This form is then translated along lines parallel to the sides of the original triangle to fill the plane. Note that the "empty" spaces are themselves the same form.

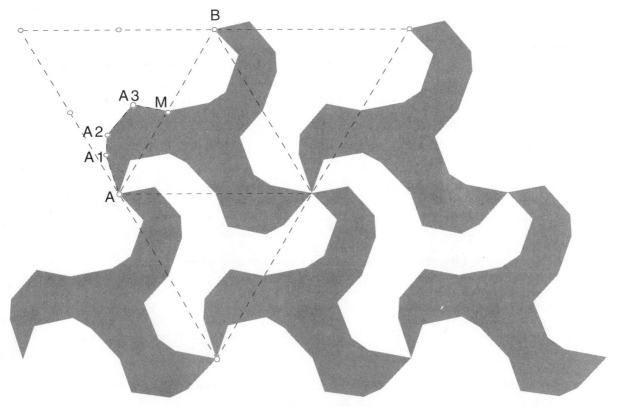

Figure 3.10

```
   File  Edit  Display  Construct  Transform  Measure  Graph  Work  Fax          ?

                                   gsp30

  ▸
  •
  ⊕        ○ side of triangle
  ╱
 ✍ᴬ

  ?



  Select & Translate
```

Focus

A curve is defined from the endpoint to the midpoint of one side of an equilateral triangle. This curve is extended over the second half of the side using a glide reflection. When the curve is replicated on the other sides of the triangle, the resulting object tessellates the plane.

Tasks

1. Open the file gsp30 and vary the segment labeled "side of triangle." Next, move one or more of the marked vertices along the perimeter of the propeller-like object. Do the two actions appear to be linked or independent of one another?

2. Create your own version of this file, using more points to define the curve. Sketch the result.

TESSELLATIONS BASED ON ROTATIONS & EQUILATERAL TRIANGLES

Tool(s) Geometers Sketchpad   Data File(s)   gsp31, gsp32   (Mac & PC)

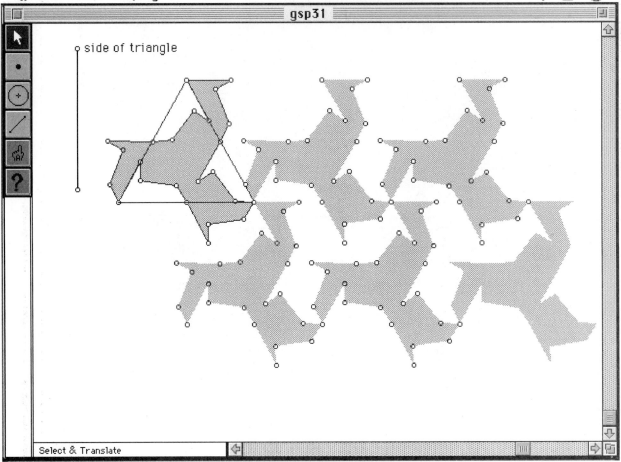

Focus

A curve is defined from the endpoint to the midpoint of one side of an equilateral triangle. This curve is extended over the second half of the side using a rotation. When the curve is replicated on the other sides of the triangle, the resulting object tessellates the plane.

Tasks

1. Open the file gsp31 and vary the segment labeled "side of triangle" and the positions of one or more of the marked vertices along the perimeter of the propeller-like object.   Measure and compare the areas and perimeters of the objects and the "negative spaces" between them. State a conjecture based on your observations. To prove your conjecture false, what sort of counter example would be necessary?

2. Repeat these activities using file gsp32.

TESSELLATIONS BASED ON ROTATIONS & SQUARES

Tool(s) Geometers Sketchpad · Data File(s) gsp33 (Mac & PC)

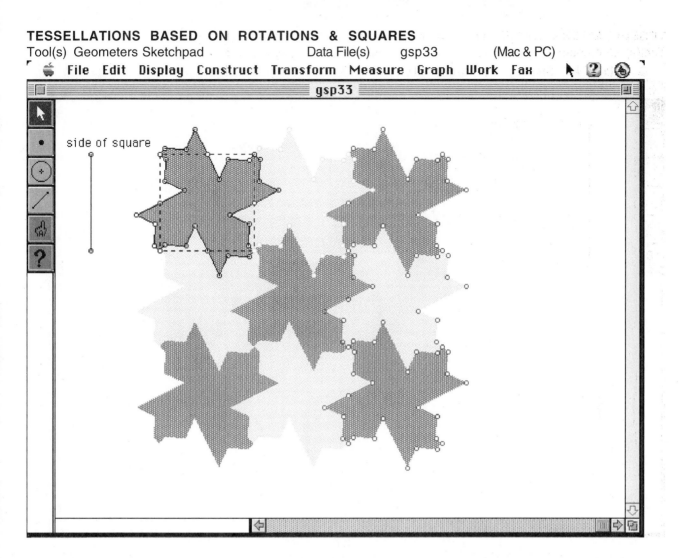

Focus

A curve is defined from the endpoint to the midpoint of one side of a square. This curve is extended over the second half of the side using a rotation. When the curve is replicated on the other sides of the square, the resulting object tessellates the plane.

Tasks

1. Open the file gsp33 and vary the segment labeled "side of square" and the positions of one or more of the marked vertices along the perimeter of the object. Measure and compare the areas and perimeters of the objects. State a conjecture based on your observations. To prove your conjecture false, what sort of counter example would be necessary?

2. Create your own version of this file, using more points to define the curve.

TESSELLATIONS BASED ON ROTATIONS & QUADRILATERALS

Tool(s) Geometers Sketchpad   Data File(s)   gsp34   (Mac & PC)

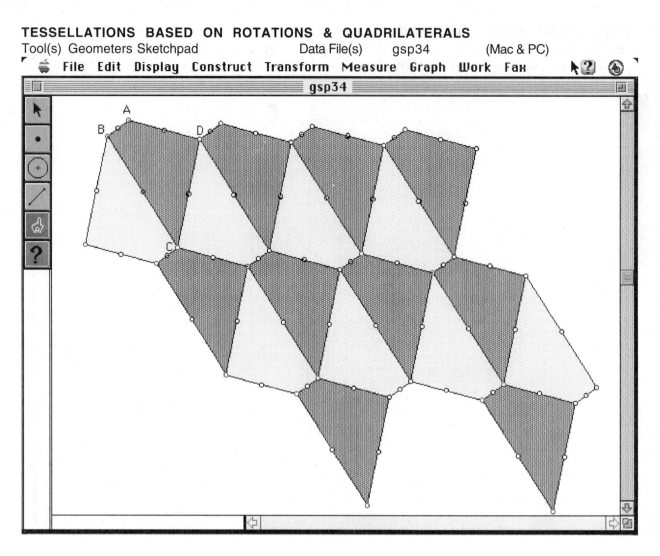

Focus

Any quadrilateral ABCD will tessellate the plane. Adjoining objects have the same side length but are rotated 180 degrees about their segment midpoints.

Tasks

1. Move vertices A, B, C, and D. Describe the effect on the tessellation.

2. Explain why any quadrilateral will tessellate the plane.

3. Argue that any two triangles with one pair of congruent sides will tessellate the plane.

TESSELLATIONS BASED ON PARALLELOGRAMS & GLIDE REFLECTIONS

Tool(s) Geometers Sketchpad   Data File(s)   gsp35, gsp36   (Mac & PC)

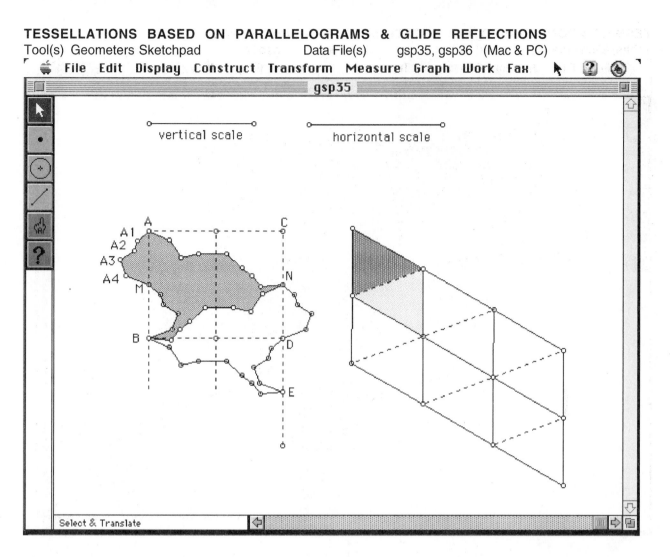

Focus

A curve is defined from the endpoint, A, to the midpoint, M, of side AB of parallelogram ABEN. This curve is extended over the second half of the side, MB, using a glide reflection. The entire curve from A to B is then reflected in the perpendicular bisector of BD and translated downward to connect N and E. A second curve is drawn from A to N and translated downward to connect B and E. The same curve is reflected in the perpendicular bisector of BD and translated downward to connect N and B. The resulting object tessellates the plane.

Tasks

1. The original figure was designed to suggest puppies. Adjust the vertices and scales to suggest a cat or some other creature.

2. Open file gsp36 and measure the area and perimeter of the two "puppies" with vertices around their boundaries. Are they the same size and shape? Discuss your findings.

TESSELLATING BABOONS

Tool(s) Geometers Sketchpad   Data File(s)   gsp37   (Mac & PC)

Focus

By adding lines and other features to a tessellation, the intended association may be enhanced.

Tasks

1. The tessellation shown above is intended to suggest an aggressive male baboon. Print a hard copy and sketch in body features that enhance the tessellation.

2. Returning to the Sketchpad, try to incorporate some of the body features from your sketch into the file gsp37. Print a hard copy of your final result.

KALI, AN INTERACTIVE 2-D EUCLIDEAN SYMMETRY PATTERN EDITOR

Tool(s) Kali   Data File(s)   none   (Mac & PC)

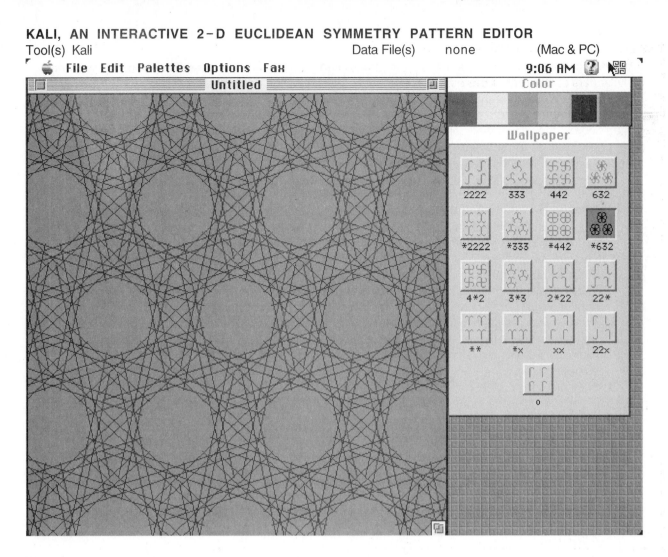

Focus

Just as there are only 7 symmetry patterns with which to create a frieze, there are only 17 symmetry patterns with which the plane may be covered. Kali provides an interactive environment in which these concepts may be explored.

Tasks

1. Using Netscape Navigator or Microsoft Internet Explorer, open the URL http://www.geom.umn.edu/software/download/kali.html and download Kali.

2. Create a design based on each of the 7 frieze symmetry patterns

3. Create a design based on each of the 17 wallpaper symmetry patterns.

GEOMETRIC PATTERNS IN ISLAMIC ART

In the *Koran*, the holy book of Islam, there are strong prohibitions against idolatry of any sort. Many of the early leaders of Islam interpreted these statements as an injunction against the representation of humans or animals in art. While this tradition may have frustrated some Islamic artists, others took up the challenge and became the greatest pattern makers of their time. Instead of covering buildings and other surfaces with human figures, they developed complex, decorative patterns with which to adorn palaces and mosques and other public places. Three types of patterns were developed: designs derived from plant life, designs based on calligraphy, and designs based on geometric shapes.

Islamic patterns emphasize five basic design principles (Norman & Stahl, 1979).

1. They are made up of a small number of repeated geometric elements that create a complex whole by repeating a few elements.
2. They are two-dimensional both in form and intent. No attempt is made to create a three-dimensional effect or representation.
3. They radiate symmetrically from a central point. Westerners who attempt to "read" the art from left to right and top to bottom often fail to see the whole work as intended by the artist.
4. They are not designed to fit within a rectangular frame. Unlike Western art which typically begins with a rectangular frame that is as much a factor in the design of the work as any other element, Islamic art begins with a central point and patterns that radiate from that point until they encounter some barrier.
5. They are constructed from patterns of circles.

HEXAGON PATTERN

Tool(s) Geometers Sketchpad   Data File(s)   gsp38   (Mac & PC)

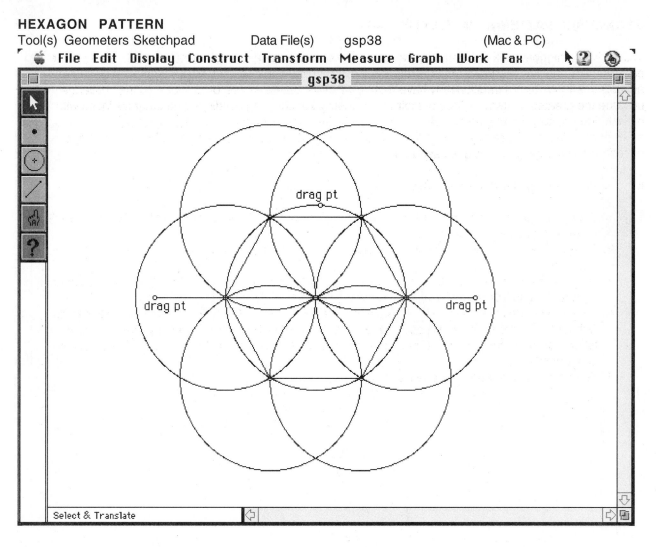

Focus

Using circles as a tool, a hexagon may be constructed without resorting to measurement.

Tasks

1. Measure the sides of the hexagon. Are they all the same? Measure the angles of the hexagon. Are they all the same?

2. Surround the original hexagon with hexagons of the same size. Describe your strategy and list the distances and angles associated with the translations.

3. Do you think that the plane could be tiled with the hexagon? Explain your reasoning.

Tool(s) Geometers Sketchpad   Data File(s)   gsp39   (Mac & PC)

Focus

Using circles as a tool, a triangle may be constructed without resorting to measurement.

Tasks

1. Measure the sides of the triangle. Are they all the same? Measure the angles of the triangle. Are they all the same?

2. Surround the original triangle with a triangle of the same size. Describe your strategy and list the distances and angles associated with the translations.

3. Do you think that the plane could be tiled with the triangle? Explain your reasoning.

Tool(s) Geometers Sketchpad   Data File(s)   gsp40   (Mac & PC)

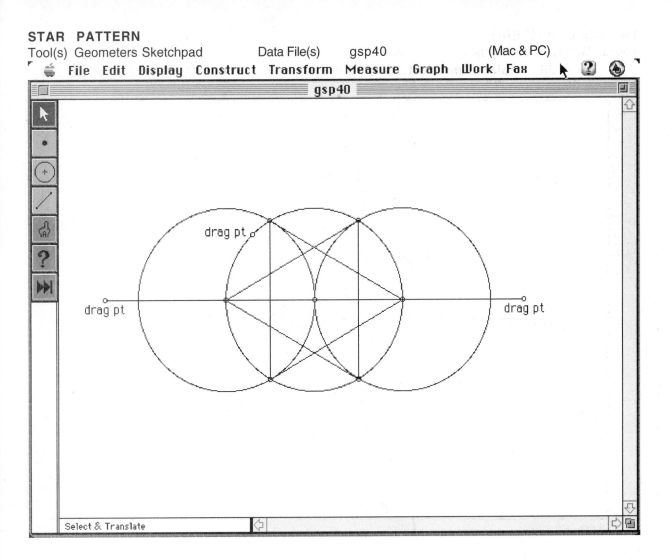

Focus

Using circles as a tool, a star may be constructed without resorting to measurement.

Tasks

1. Measure the sides of the star. Are they all the same? Measure the angles of the star. Are they all the same?

2. Do you think that the plane could be tiled with the star? Explain your reasoning.

SQUARE PATTERN

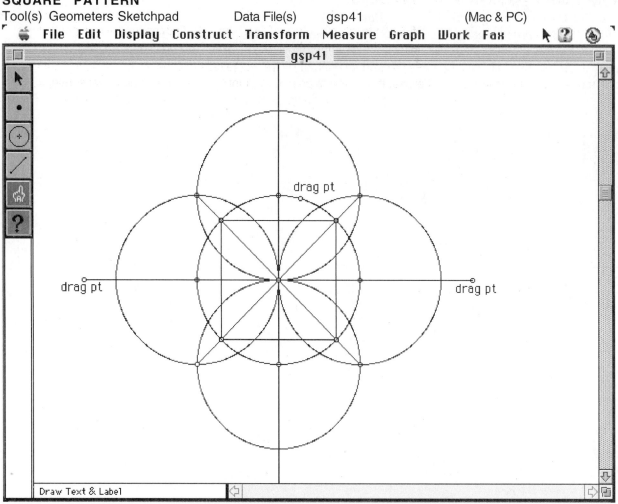

Focus

Using circles as a tool, a square may be constructed without resorting to measurement.

Tasks

1. Measure the sides of the square. Are they all the same?   Measure the angles of the square. Are they all the same?

2. Surround the original square with a triangle of the same size. Describe your strategy and list the distances and angles associated with the translations.

3. Do you think that the plane could be tiled with the square? Explain your reasoning.

THE TRANSFORMATION OF INVERSION

The transformation of inversion is defined in terms of a circle of inversion. To find the image of a point P under inversion, a line is drawn containing the center of the circle of inversion and point P. The image of point P lies on this line a distance r^2 / OP from the center on the same side of O as P. In the figure below, the image of point P is point Q. Likewise, the image of point Q is point P. In a purely numerical sense, inversion is about two distances that always have the same product. The geometric interpretations, however, are much more interesting.

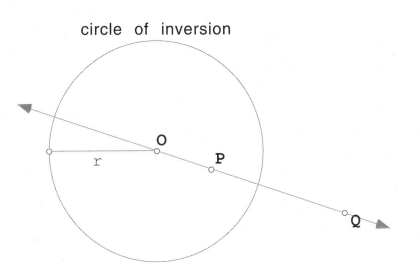

Figure 3.11

THE IMAGE OF A POINT UNDER INVERSION

Tool(s) Geometers Sketchpad   Data File(s)   gsp42   (Mac & PC)

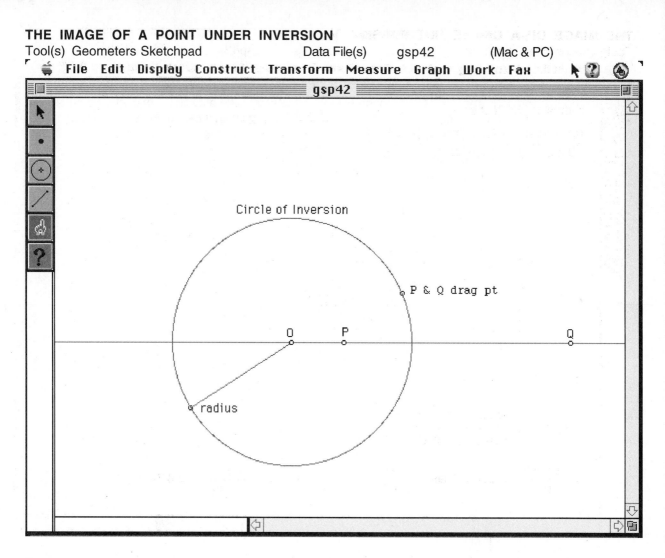

Focus

Two points P and Q are inverse points with respect to a point O and a given positive real number r if $OP * OQ = r^2$, and O, P, Q are collinear, where r is the radius of inversion.

Tasks

1. Use the measure menu to find the distances from O to Q, O to P, and the length of the radius of inversion.

2. Use the Calculate option in the Measure pull-down menu to find OP*OQ and the square of the radius of inversion.

3. Finally, move the "P&Q drag pt." around the circle, changing the distances of P and Q from O. Discuss your observations.

THE IMAGE OF A CIRCLE NOT PASSING THROUGH THE CENTER OF INVERSION

Tool(s) Geometers Sketchpad   Data File(s)   gsp43   (Mac & PC)

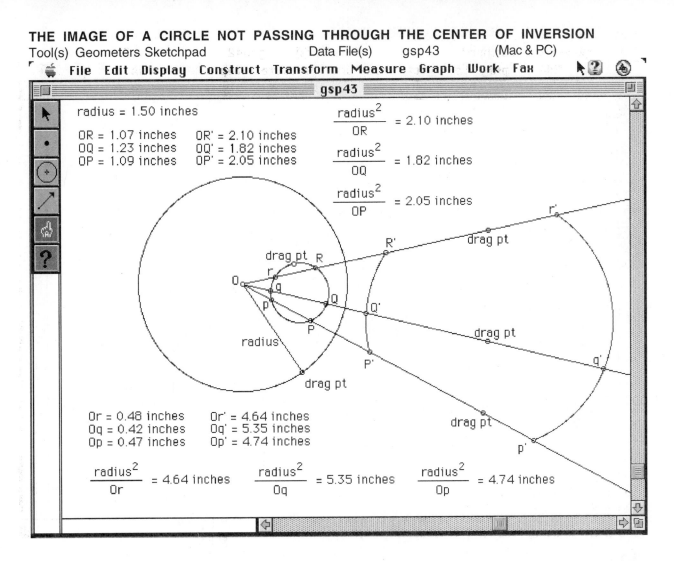

Focus

The image of a circle not passing through the center of inversion.

Tasks

1. Experiment with the drag pts and observe the effects they have on the positioning of objects and images and the calculations displayed. To obtain the correct image following any movement of the drag pts, you must move points R', r', P', p', Q', and q' so that their distances from O match their calculated positions. When you have correctly done so, what shape does the image appear to be?

2. Could you view the outside circle as the object and the inside circle as the image in this process? Would the construction permit you to reverse the roles of object and image in this way? Why or why not?

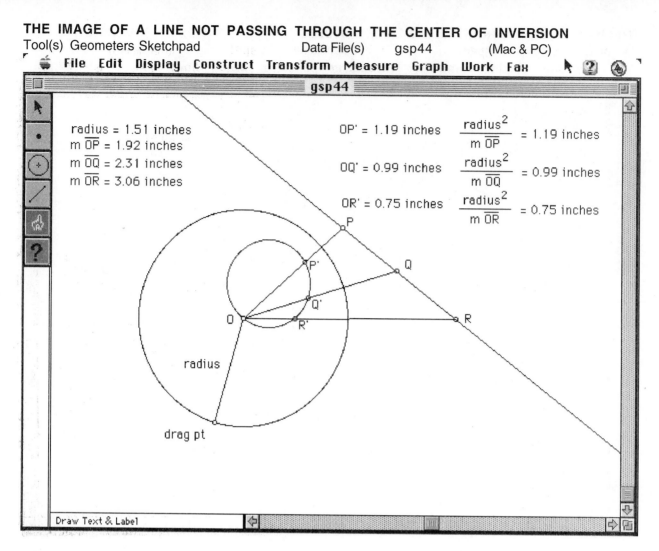

Focus

The image of a line not passing through the center of inversion.

Tasks

1. Experiment with the drag pt and points P, Q, and R. To obtain the correct image following any movement of the drag pts, you must move points P´, Q´, and R´ so that their distances from O match their calculated positions. When you have correctly done so, what shape does the image appear to be?

2. Where is the point whose image is O, the center of inversion?

INVERSE POINTS AND ORTHOGONAL CIRCLES

Tool(s) Geometers Sketchpad   Data File(s)   gsp45   (Mac & PC)

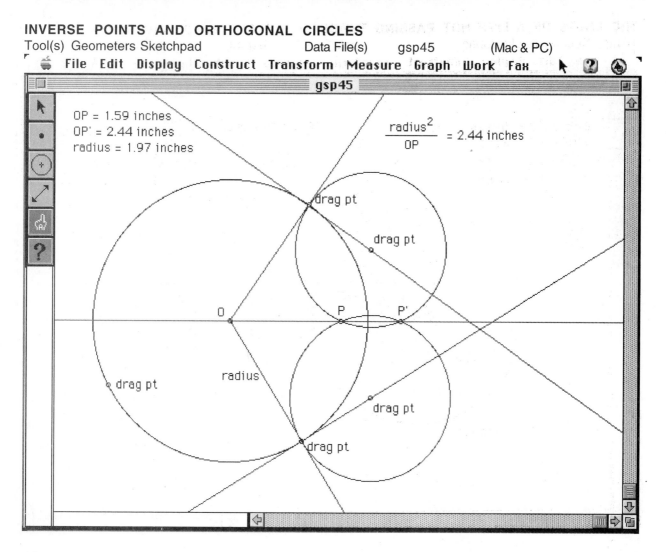

Focus

The intersection points of two circles, each orthogonal to the circle of inversion, are inverse points relative to the circle of inversion.

Tasks

1. Explain the basis for the construction of the orthogonal circles shown above.

2. Move the drag pts and observe the range of motions available to P and P´. Under which circumstances do you obtain the widest range of separations between P and P'?

FINDING A CIRCLE OF INVERSION GIVEN TWO POINTS

Tool(s) Geometers Sketchpad   Data File(s)   gsp46   (Mac & PC)

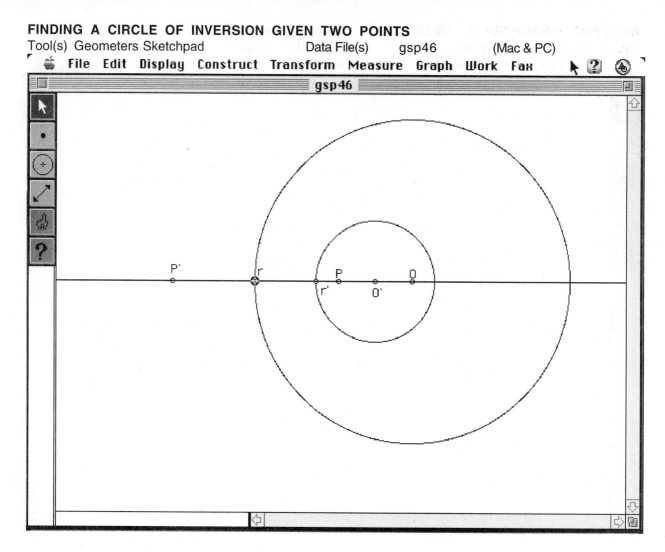

Focus

Given two points, it is possible to find more than one circle of inversion for which the two points are inverses of one another.

Tasks

1. Given points P and P′, find two circles with centers O and O′ and radii r and r′ respectively for which OP * OP′ = r² and O′P * O′P′ = r′².

2. Explore methods for finding circles of inversion involving more than "guess and check" strategies. Discuss your findings.

Tool(s) Geometers Sketchpad   Data File(s)   gsp47   (Mac & PC)

File   Edit   Display   Construct   Transform   Measure   Graph   Work   Fax

Focus

The image of a straight line not through the center of inversion, under the transformation of inversion, is a circle passing through the center of inversion. The equations for the inversion transformation of a point $P(x,y)$ into $P'(x',y')$ relative to the circle centered at the origin are:

$X' = XR^2 / (X^2 + Y^2)$ and $Y' = YR^2 / (X^2 + Y^2)$.

Tasks

1. Find the coordinates of P1, P2, and P3 with coordinate tool under the measure menu.

2. Use the Calculate option in the Measure pull-down menu to find $P1'$, $P2'$, and $P3'$.

3. Use the Plot tool in the Graph pull-down menu to plot these new points.

4. Use the Arc Thru 3 Points tool in the Construct pull-down menu to plot the image circle.

CHAPTER 4

PROJECTIVE GEOMETRY

The geometry of Euclid is based on measurement. Early in the nineteenth century, a different sort of geometry was developed called "projective geometry" in which measurement played no part. The origins of projective geometry may be seen in the attempts of Renaissance artists and architects to achieve realism in their drawings and paintings of three-dimensional objects on two-dimensional surfaces. These artists and architects sought and found a mathematical basis for perspective drawing.

The first written exposition on how to use geometrical ideas about one-point perspective to make paintings more realistic was published in Florence, Italy, in 1435 by Leon Battista Alberti. The idea of perspective was first introduced in the study of optics, also known as the science of perspective. While Euclid's treatise on optics contributed greatly to Alberti's thinking, it did not encompass all that was necessary to meet the needs of artists. Euclid neglected to answer several questions, such as why parallel lines appear to converge, or what causes things in the distance to appear lighter than things in foreground.

Over the past 500 years, work has been done to refine linear perspective. Artists, architects, and engineers today still apply the basic ideas expressed by Alberti to recreate three-dimensional objects on two-dimensional planes like television screens and computer monitors. Although credit is give to Brunellschi for the development of linear perspective, we owe Leon Battista Alberti for it's written record.

The activities in this chapter explore the idea of linear perspective and introduce a number of technologies related to the scientific and creative uses of projective geometry.

VANISHING POINTS & HORIZON LINES

Tool(s) Geometers Sketchpad   Data File(s)   gsp48   (Mac & PC)

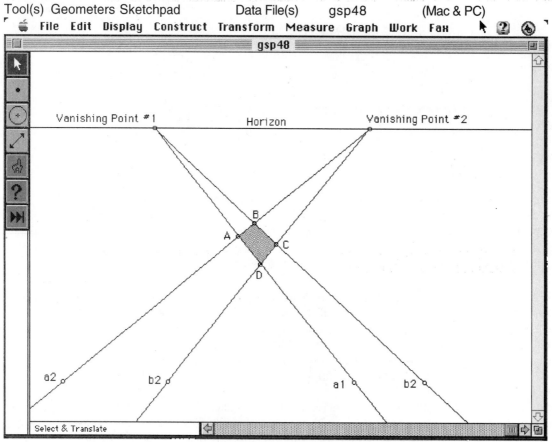

Focus

Parallel lines in the plane appear to converge at a distant point on the horizon.

Tasks

1. Move Vanishing Point #1 and Vanishing Point #2 along the horizon line. How do these motions appear to affect the position and size of the shaded quadrilateral relative to you?

2. Move the Horizon up and down the screen. How do these motions appear to affect the position and size of the shaded quadrilateral relative to you?

3. Change the separation between the lines by dragging points a1, b1, a2, and b2. How do these motions appear to affect the position and size of the shaded quadrilateral relative to you?

4. Position the vanishing points, horizon, and lines to create the impression that the shaded quadrilateral is a rectangle close to your feet.

DISTANCE LINES
Tool(s) Geometers Sketchpad   Data File(s)   gsp49   (Mac & PC)

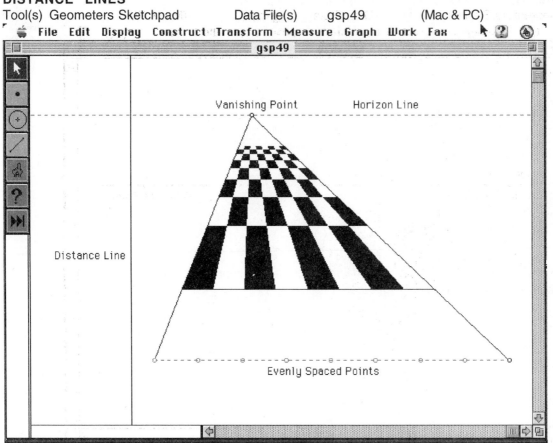

Focus

> The apparent distance from the observer to the object is variable, as is the apparent height of the observer above the plane.

Tasks

> 1. Vary the position of the Distance Line. How does this affect the shape and position of the object? Your height above the plane?

> 2. Vary the position of the Horizon. How does this affect the shape and position of the object? Your height above the plane?

> 3. Vary the position of the Vanishing Point. How does this affect the shape and position of the object? Your height above the plane?

3-D CUBE

Tool(s) Geometers Sketchpad   Data File(s)   gsp50   (Mac & PC)

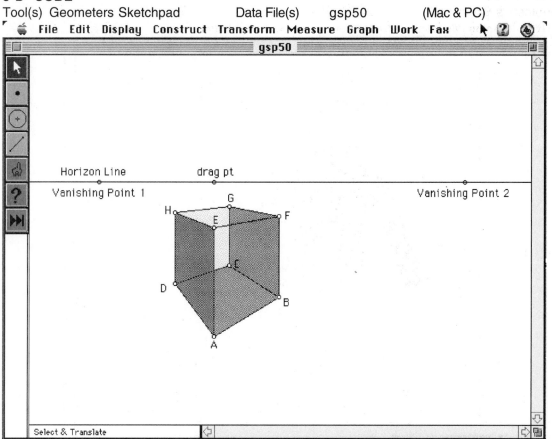

Focus

A dynamic, 2-point perspective representation of a cube.

Tasks

1. Move Vanishing Point 1 and Vanishing Point 2 along the Horizon Line.   Move the drag pt along the Horizon Line. Describe the apparent effect on the shape and position of the object.

2. Drag point E above the Horizon Line. Describe the apparent effect on the shape and position of the object.

3. Return point E to its original position and drag point A above the Horizon Line. Describe the apparent effect on the shape and position of the object.

3-D TETRAHEDRON

Tool(s) Geometers Sketchpad   Data File(s)   gsp51   (Mac & PC)

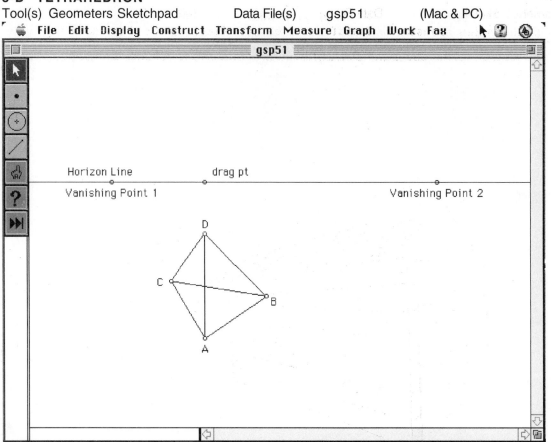

Focus

A dynamic, 2-point perspective representation of a tetrahedron.

Tasks

1. Color and shade the faces of the tetrahedron to better reveal its features. What choice of color and shadings do you find most helpful?

2. Drag point A above the Horizon Line. Describe the apparent effect on the shape and position of the tetrahedron. How does this affect what you can see of the object?

3. Modify gsp50 and/or gsp51 to create a 3-D perspective drawing of an object other than a cube or a tetrahedron.

3-D HOUSE

Tool(s) Geometers Sketchpad   Data File(s)   gsp52   (Mac & PC)

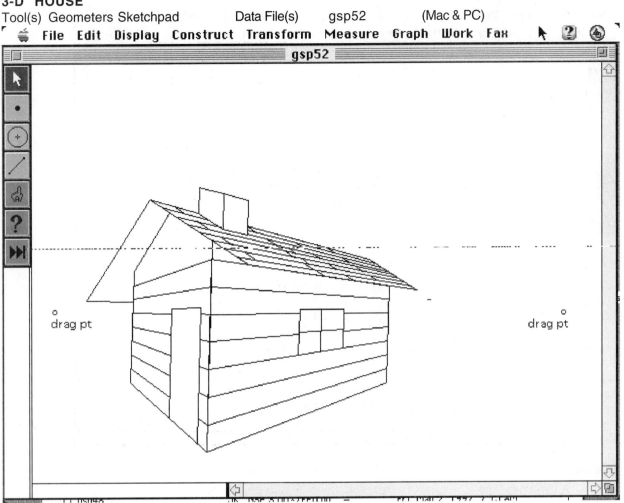

Focus

A dynamic, 2-point perspective representation of a house.

Tasks

1. Vary the positions of the drag pts. Describe how the positions of the drag pts affect the shape and location of the house relative to your viewpoint.

2. Color and shade the house to better reveal its features.

TWO VIEWS

Tool(s) Geometers Sketchpad Data File(s)   gsp53, gsp53.1, gsp53.2 (Mac & PC)

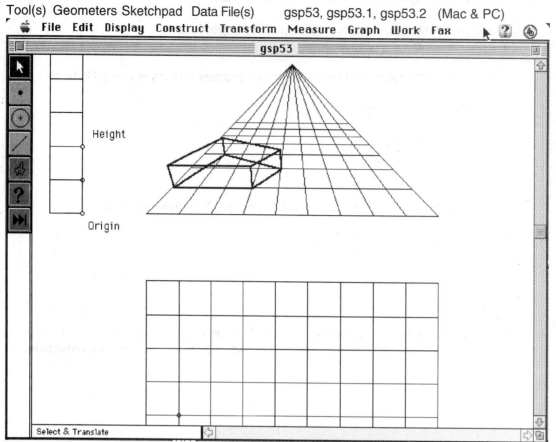

Focus

A 1-point perspective view of a right prism.

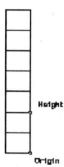

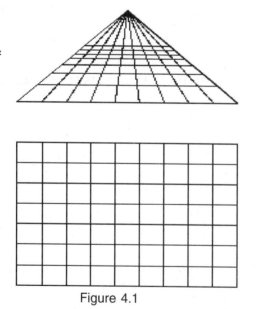

Figure 4.1

Figure 4.2

Open the files gsp53.1 (Figure 4.1) and gsp53.2 (Figure 4.2).

Holding down the shift key, select six points as specified in gsp53.2.
- The point marked Origin in gsp53.1.

- Four points on the rectangular grid representing consecutive vertices of a quadrilateral.

- The point marked Height in gsp53.1.

Click the Fast button in gsp53.2.

Tasks

1. Describe the features of the prism as represented in the 1-point perspective view.

2. Repeat the procedure without deleting the first prism. The second prism should be shorter than the first and positioned so as not to intersect the first prism. Color one prism red and the other blue.

3. Create a nested series of square prisms with bases of length 5, 4, and 3 units and heights of 1, 2, and 3 units, respectively. Color and shade the prisms to better reveal their features and relationships to one another.

4. Create an object shaped like a set of stairs having four steps. Color and shade the result to better reveal the overall shape of the stairs.

ALBERTI'S METHOD

Tool(s) Geometers Sketchpad   Data File(s) gsp54   (Mac & PC)

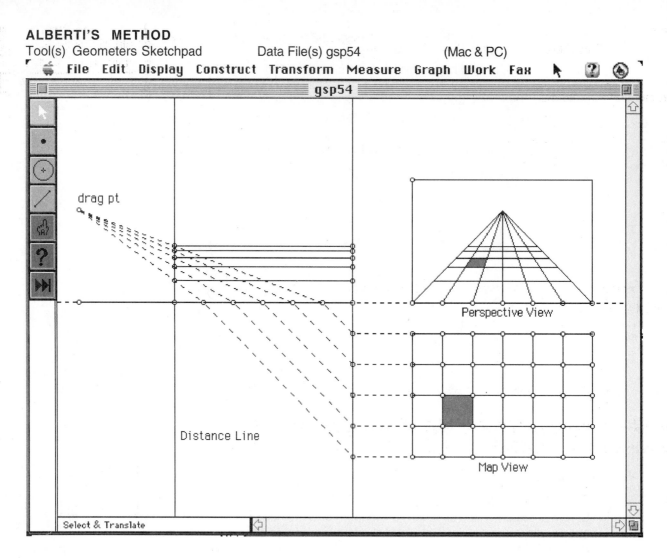

Focus

Alberti's method for creating a 1-point perspective view.

Tasks

1. Vary the position of the Distance Line.   How does this affect the shape and position of the object? Your height above the plane?

2. Vary the position of the drag pt. How does this affect the shape and position of the object? Your height above the plane?

3. This procedure creates an illusion of depth. Is that illusion true to reality or is it a false view of the world?

Dividing Space 1

Tool(s) Geometers Sketchpad   Data File(s)   gsp57   (Mac & PC)

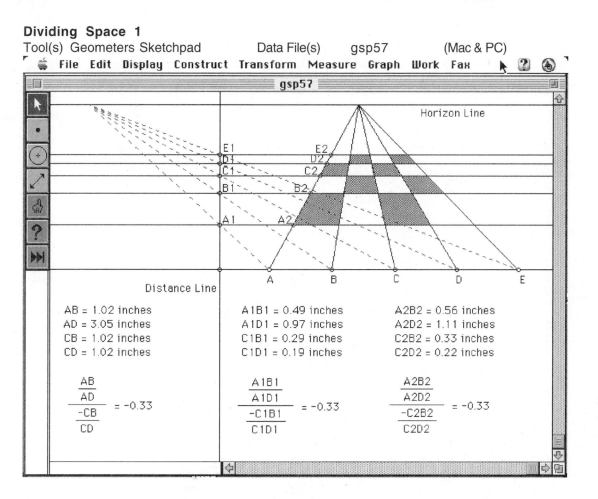

Focus

The cross ratio may be used as a indicator of how an object divides space.   In this activity, the cross ratio is used to examine Alberti's Method.

Tasks

1. Describe the three sets of points for which the cross ratio is computed and their relationships to one another.

2. Vary the position of the Distance Line. Vary the position of the Horizon Line. Describe the effect of these actions on the cross ratios.

3. Do these observations strengthen Alberti's assertion that his method creates a true view of the world? If so, how? If not, why not?

DIVIDING SPACE 2

Tool(s) Geometers Sketchpad   Data File(s)   gsp55   (Mac & PC)

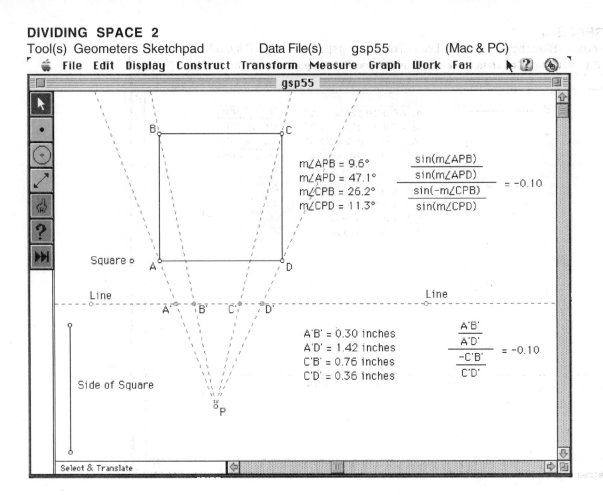

Focus

The cross ratio may be computed on the basis of either angular or linear measurements and used as an indicator of how an object divides the field of view of an observer. In gsp55, the observer is at point P and the object is square ABCD.

Tasks

1. Describe the effect on the cross ratio of moving the line. Of resizing the square using the Side of Square segment. Of rotating the square using the Square drag point.

2. Vary the position of point P. Describe the effect on the cross ratio.

3. Do both methods of computing the cross ratio always return the same result?

DIVIDING SPACE 3

Tool(s) Geometers Sketchpad   Data File(s)   gsp56   (Mac & PC)

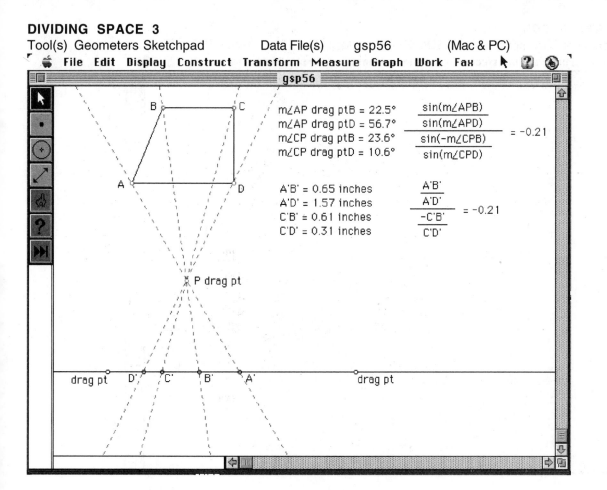

Focus

A comparison of the cross ratios of different quadrilaterals.

Tasks

1. Reposition points A, B, C, and D, creating a different quadrilateral. What happens to the cross ratio?

2. Is it possible for different quadrilaterals to have the same cross ratio relative to some observer? If so, under what circumstances?

3. Discuss the implications of your observations for the following situation: You are standing at point P on a large plot of land, drawing a map showing the location of buildings and other easily recognized objects.

PAPER STRIP TECHNIQUE

Tool(s) Geometers Sketchpad   Data File(s)   gsp58   (Mac & PC)

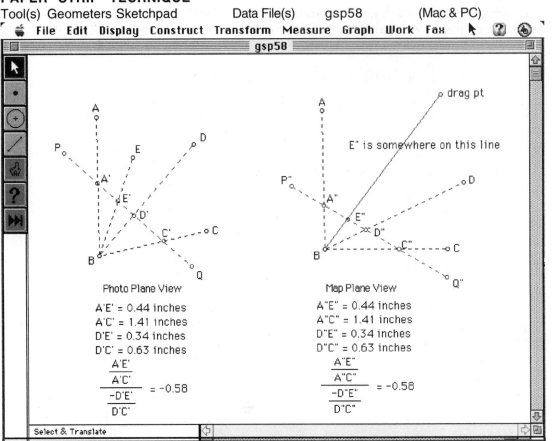

Focus

The paper strip technique is a map-maker's trick for converting survey data into map data. In the Photo Plane View shown above, points A, B, C, D, and E are marked. In the Map Plane View, only points A, B, C, and D are seen. The task is to use the data in the Photo Plane Image to construct the location of point E in the Map Plane Image.

1. Point B in the Photo Plane View is selected as a reference point. Rays are drawn from B through points A, C, D, and E. A line PQ is drawn intersecting these rays in points A´, C´, D´, and E´. You may verify that the cross ratio is independent of your choice of lines by moving points P and Q.
2. Point B in the Map Plane View is selected as a reference point. Rays are drawn from B through points A, C, and D. A line P´´Q´´ is drawn intersecting these rays in points A´´, C´´, and D´´. You may verify that the cross ratio is independent of your choice of lines by moving points P´´ and Q´´.
3. The drag pt in the Map Plane View is moved until the cross ratio is the same as the cross ration computed in the Photo Plane. Point E´´ is then known to lie somewhere on the line from point B to the drag pt. In the paper strip technique, the cross ratio is not computed. Instead, a strip of paper is laid along line PQ and the edge of the paper is marked at the points of intersection with the rays. This strip is then transferred to the Map Plane View and the marks aligned with the rays from point B to points A, C, and D. A mark is made on the Map Plane View corresponding to the point on the paper strip associated with point E. Point E is then known to lie on the ray from point B through that mark.

92

Tasks

1. Repeat the procedure using point C as a reference point in both Views. This will produce a second line known to contain point E in the Map Plane View. Point E lies at the intersection of those two lines.

2. Use the paper strip version of this procedure on the points shown below.

°B   °B

°A   °E   °A   °C

                    °C

        °D   °D

    Photo Plane   Map Plane

3. Discuss the basis for this procedure, remembering that the cross ratio may be computed by either linear or angular measurements and that the results are always the same for a given set of points.

Figure 4.3

The realistic graphic shown in Figure 4.3 was created using digital elevation data (DEM) of the Mt. McKinley region of Alaska from the US Geological Survey (USGS) (http://edcwww.cr.usgs.gov/glis/hyper/guide/1_dgr_demfig/states/) and VistaPro, a landscape-rendering program from ROMTECH, Inc. (http://www.romt.com/). After specifying a camera position and field of view, VistaPro rendered the landscape using projective and fractal geometry. As a result, the shape of the land is "correct" but all surface details are artificial. VistaPro also may be used to create landscapes based entirely on fractal geometry, that is, from mathematical models rather than USGS data. The mountainous terrain in Figure 4.4 was created in this manner.

Figure 4.4

DOWNLOADING VISTAPRO
Tool(s) Netscape Navigator or Microsoft Internet Explorer   Data File(s) none (Mac & PC)

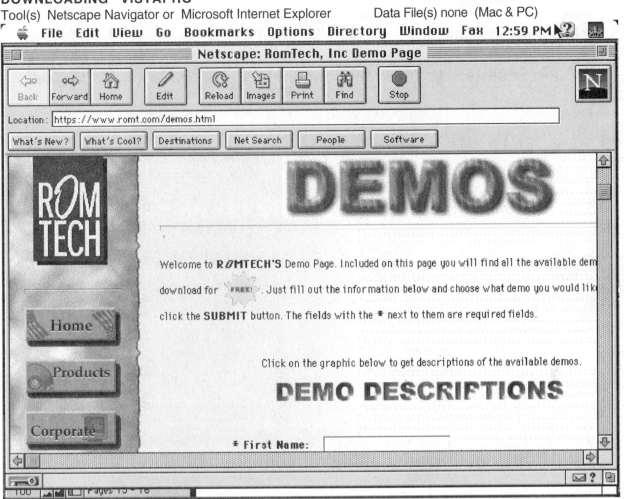

Focus

Download the demo version of VistaPro from ROMTECH's WWW site and install it on your Macintosh, DOS, or Windows computer.

Tasks

1. Using Netscape Navigator or Microsoft Internet Explorer, Open the URL http://www.romt.com/demos.html   Fill in the information requested, select the Macintosh, DOS, or Windows version of VistaPro, and submit the request.

2. The file that you receive is compressed. If your computer is configured for receiving files over the Internet, it should decompress the VistaPro file automatically. If it doesn't, try double clicking on the file icon. Sometimes, this action launches decompression software on the computer. If the file does not decompress, find a computer that will complete the task, copy the ready-to-use demo program to a disk, and carry the disk to wherever you want to work. You will probably waste less time than if you try to configure your computer without the assistance of a system operator.

Focus

Explore the options available in the VistaPro demo.

Tasks

1. Launch VistaPro by double clicking on the icon. A set of windows like that shown above will appear. The largest window is a topographic map of the El Capitan area of Yosemite National Park. What purpose do the different colors serve?

2. The first order of business is to find out exactly what VistaPro does. Select Render in the Landscape pull-down menu. After a few seconds, a rough image of Yosemite Valley should appear. Describe the result. A sampler of VistaPro images is available at the ROMTECH WWW site (http://www.romt.com/products/vista/vpimages/).

3. A number of icons appear to the left of and below the topographic map. Double click on each icon. As you do so, a window will open containing information and/or offering you the opportunity to set parameters and other features available in VistaPro. Write a brief description of the purpose and/or function of each icon:

X: 3660	ΔX: 3150
Y: 3000	ΔY: -1830
	ΔZ: 178
Z: 1013	ΔR: 3647

4. Select Fractal in the Windows pull-down menu. Click on the Random button to create a fractal landscape. Render the fractal landscape. How realistic is the fractal landscape compared to your rendering of Yosemite Valley?

VIRTUAL REALITY ORIENTATION
Tool(s) Netscape Navigator or Microsoft Internet Explorer   Data File(s) none (Mac & PC)

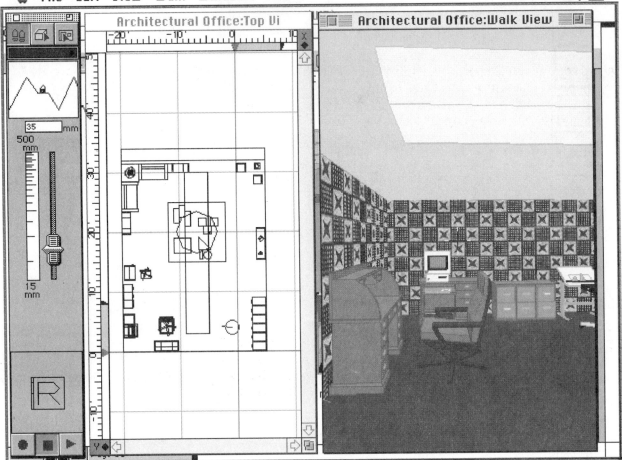

Focus
In addition to rendering realistic landscapes, virtual reality extends to the representation and exploration of man-made environments.

Tasks
1. Using Netscape Navigator or Microsoft Internet Explorer, open the URL http://www.virtus.com/products_downloaddemo.html and download the demo version of Virtus Corporation's WalkThrough Pro suitable for your computer.

2. Print out the files in the WalkThrough Pro TEXT folder for reference.

3. Under the File pull-down menu, select Open. When the dialog box opens, open the Models folder and select the file Architectural Office.

4. Click the arrow in the lower left portion of the screen to start a tour of the Architectural Office. Note the relationship between the walk (projective) view and the schematic, top view. Where are you, the observer, in the top view?

5. At some point in the tour, stop the action by clicking on the square to the left of the arrow. Move the pointer to your position and drag the observer icon to a new position in the top view. Click in the walk view window and hold down the mouse key in various positions relative to the cross hairs in the center of the window. Describe what happens.

6. Move the slider on the left side of the screen. Describe what happens.

7. Select Library in the File pull-down menu. When the dialog box opens, select Sample Library. Using Copy and Paste in the Edit window, transfer a microwave into the Architectural Office and place it on the stool in front of the drafting table. Use the Change View options in the View pull-down menu to help you drag-and-drop the microwave into position.

8. Create a New file in which you design a college dorm room. You will need to do so in one setting, as the demo version of WalkThrough Pro does not save files for later use. When you have completed the design, take a snapshot of the computer screen and print it out.

9. What value do you see in making virtual reality tools available to students? How would you approach doing so in a geometry class?

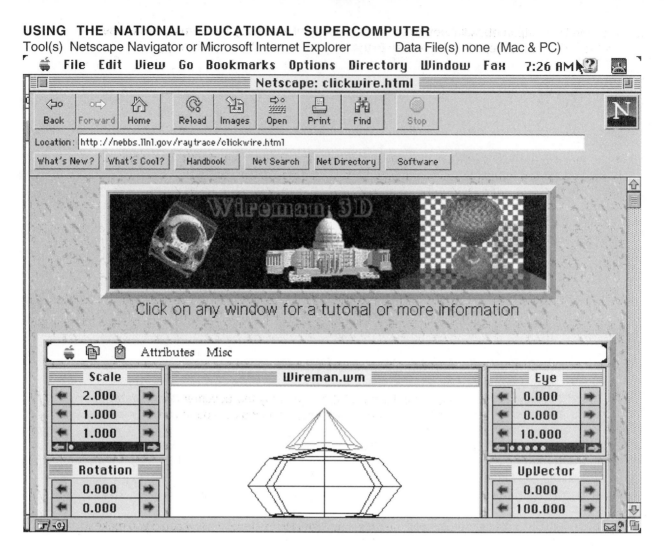

Focus

The National Educational Supercomputer Program at Lawrence Livermore National Laboratory provides students and teachers access to high performance tools for creating realistic, ray-traced movies. Using a client program called Wireman on a PC or Macintosh computer, the user creates a series of perspective views of one or more objects. The file containing the frames is transferred to the National Educational Supercomputer where it is submitted to a Cray supercomputer for processing. When the processing is complete, the user downloads the finished movie to a PC or Macintosh computer for viewing. (These instructions assume a Macintosh user. PC users will notice some differences.)

Tasks

1. Using Netscape Navigator or Microsoft Internet Explorer, open the URL http://nebbs.llnl.gov/DP.html and study the diagram explaining distributed processing. Write a paragraph summarizing the information in the diagram.

2. Open the URL http://nebbs.llnl.gov/exercise/exercise.html and study the Wireman Exercise carefully. Relate the steps in the exercise to the diagram on distributed processing.

3. Open the URL ftp://nebbs.llnl.gov/pub/NESP_Software/ and download the PC or Macintosh versions of Wireman and Movie. Macintosh users should also download Fetch.

4. Replicate the sequence of steps found in the Wireman Exercise on your computer.

5. Open the URL http://nebbs.llnl.gov/movie/movie.html and study the documentation for Movie. Once you understand the use of this tool, launch your copy and look at your movie.

6. Discuss your movie. How realistic is it? Are you satisfied with the result? How could you make it better?

7. The Movie software comes with several examples. Look at the examples and select one for discussion. Describe the objects and action in the movie and the geometric decisions and transformations that went into creating the movie.

8. Repeat the entire process, zooming in on a donut-shaped, red and white checkered object embedded in a translucent blue sphere. In the process of zooming in, pass through the hole of the donut.

CHAPTER 5

FRACTAL GEOMETRY

During the 1980s, the study of dynamical systems emerged as one of the fastest-growing branches of mathematics. Thanks to the efforts of Benoit Mandelbrot, Heinz-Otto Pietgen, Michael Barnsley, James Gleick, and a number of other researchers and writers, a fascinated public was introduced to the branch of dynamical systems called fractal geometry. Although the pedigree of fractal geometry is still being argued by professional mathematicians, the computer science community has taken this new tool and used it to dazzle the public with spectacular color graphics, including realistic animations of alien worlds.

Like many other branches of mathematics, fractal geometry may be investigated at several levels. The purpose of this chapter is to provide exploratory activities in which students may investigate the concepts of self-similarity and fractal dimension. The chapter begins by introducing the concept of deterministic, self-similar fractals using a tool familiar to many elementary and secondary school teachers, the Geometers Sketchpad. Since very few people are capable of imagining the final form of a fractal, the availability of computer modeling and visualization tools is a critical factor in bringing this type of mathematics into the classroom.

INTRODUCTION TO FRACTAL CURVES

KOCH SNOWFLAKE CURVE 1

Tool(s) Geometers Sketchpad   Data File(s) gsp59 & gsp59.1   (Mac & PC)

Focus

The Koch snowflake curve is generated using an iterative process in which the same geometrical transformation is applied over and over on smaller and smaller scales.

Tasks

1. Under the File pull-down menu, select New Sketch, then Open the file gsp59. In the New Sketch window, select two points. Switching to the gsp59 window, click on the Fast button. A Depth of Recursion dialogue box will open. Generate a sequence of curves by entering the numbers 0, 1, and 2. For each case, determine the number of segments.

2. If this process could be continued an infinite number of times, what would happen to the number of segments? To the length of the segments? To the length of the curve?

Tool(s) Geometers Sketchpad   Data File(s) gsp60   (Mac & PC)

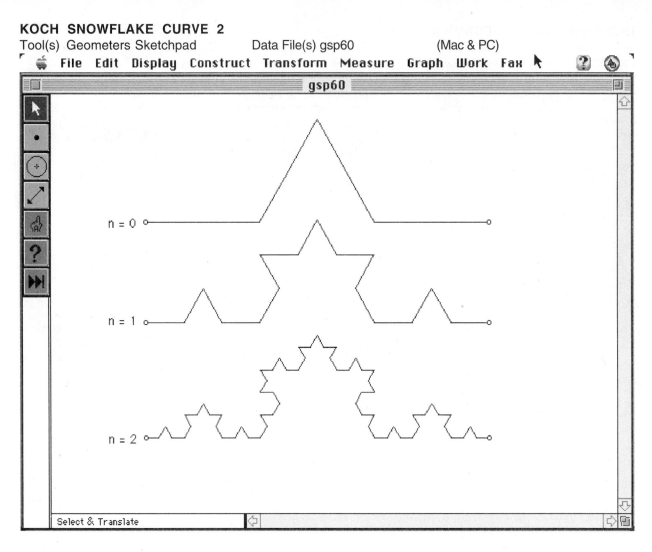

Focus

The Koch snowflake curve is self-similar, that is, small pieces resemble larger pieces.

Tasks

1. Using the point at the left end of the n = 0 curve, drag the curve down onto the n = 1 curve. How many triangles are added to the n = 0 curve to obtain the n = 1 curve?

2. Using the point at the left end of the n = 1 curve, drag the curve down onto the n = 2 curve. How many triangles are added to the n = 1 curve to obtain the n = 2 curve?

3. How many triangles would be added to the n = 2 curve to obtain the n = 3 curve? Test your answer by using gsp59 to draw the n = 3 curve between the endpoints of the n = 1 curve, then dragging the n = 3 curve down onto the n = 2 curve.

Tool(s) Geometers Sketchpad   Data File(s) gsp61   (Mac & PC)

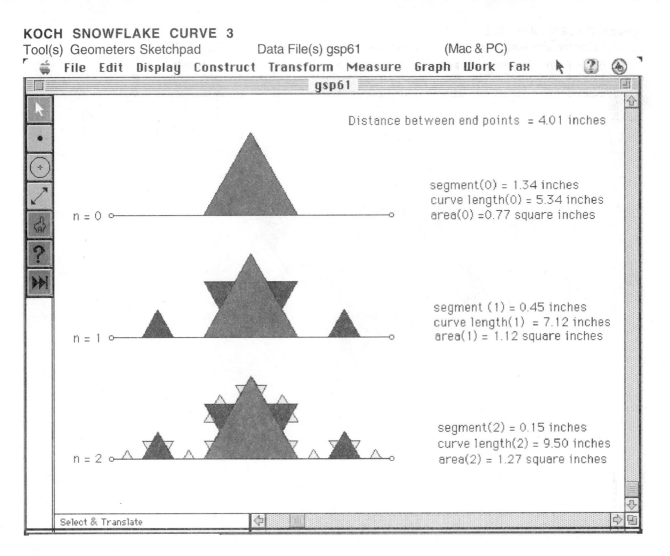

Focus

As n goes to infinity, what happens to the Koch snowflake curve length and area?

Tasks

1. Compute segment and curve lengths for n = 0, 1, and 2. Find an expression for the length of the segments in the nth iteration. For the length of the curve in the nth iteration.

2. Compare the shaded area for n = 0, 1, and 2. Find an expression for the shaded area in the nth iteration.

3. As n goes to infinity, what happens to the curve length? To the shaded area?

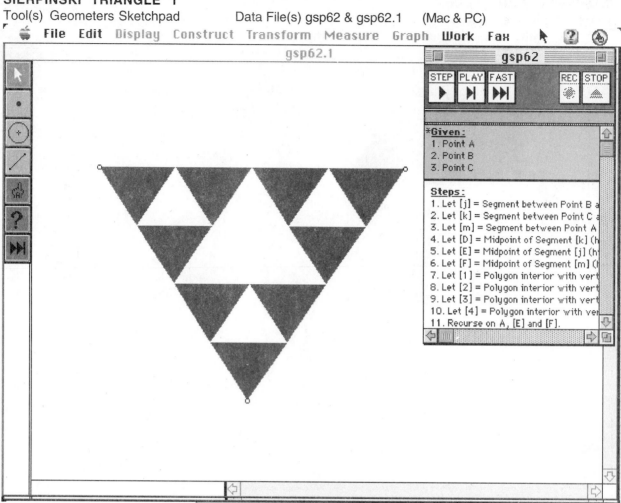

Focus

The Sierpinski triangle is generated using an iterative process in which the same geometrical transformation is applied over and over on smaller and smaller scales.

Tasks

1. Under the File pull-down menu, select New Sketch, then Open the file gsp62. In the New Sketch window, select three points. Switching to the gsp62 window, click on the Fast button. A Depth of Recursion dialogue box will open. Generate a sequence of triangles by entering the numbers 0, 1, and 2. For each case, determine the number of shaded triangles.

2. If this process could be continued an infinite number of times, what do you think would happen to the number of triangles? To the perimeter of the triangles? To the area of the triangles?

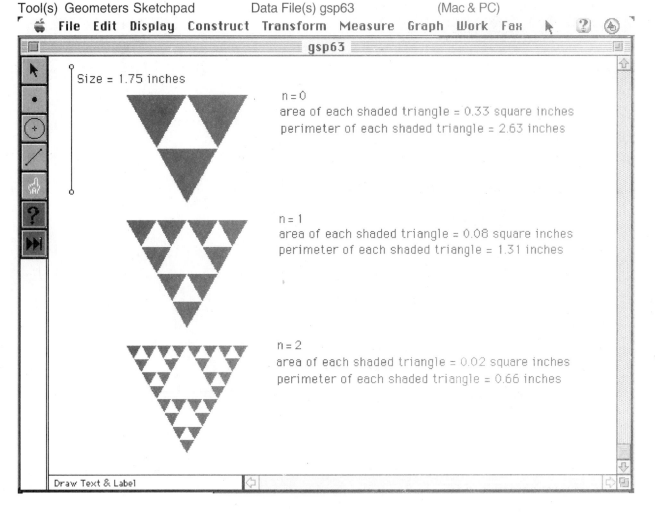

Focus

As n goes to infinity, what happens to the Sierpinski triangles' perimeters and areas?

Tasks

1. Compare the areas of each shaded triangle for n = 0, 1, and 2. Find an expression for the sum of the areas of the shaded triangles in the nth iteration.

2. Compare the perimeters of each shaded triangle for n = 0, 1, and 2. Find an expression for the sum of the perimeters of the shaded triangles in the nth iteration.

3. As n goes to infinity, what happens to the sum of the areas of the shaded triangles? To the sum of the perimeters of the shaded triangles?

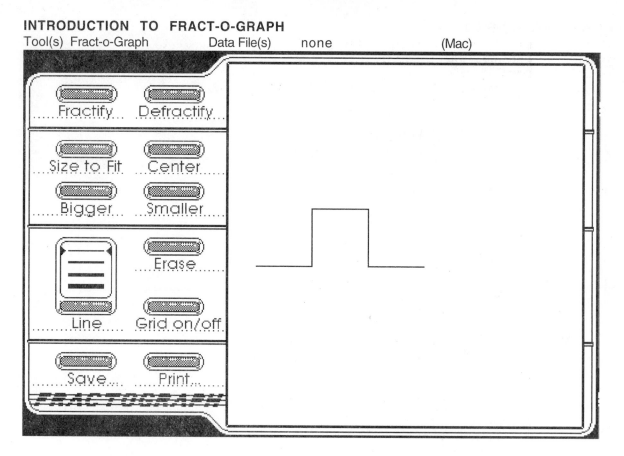

Focus

Fract-o-Graph is an easy-to-use program for creating self-similar (fractal) curves.

Tasks

1. Start Fract-o-Graph. When the control panel appears, select the line thickness that you prefer by clicking on the button at the bottom of the Line menu. Next, draw the object shown in Figure 5.1. The object is created by single clicking the mouse at the ends of each segment. Double click the last point. Once the drawing is complete, click on Size to Fit in the control panel. Click Fractify in the control panel to obtain Figure 5.2. Continue clicking to obtain Figures 5.3 and 5.4. Continue in this manner until no change is observed.

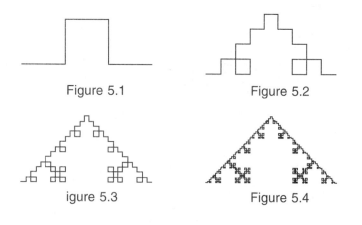

Figure 5.1   Figure 5.2

igure 5.3   Figure 5.4

2. Experiment with other starting shapes as the basis for self-similar curves. Try to find a curve that completely fills some region of the plane.

PLANE FILLING CURVES

Tool(s) Fract-o-Graph   Data File(s)   none   (Mac)

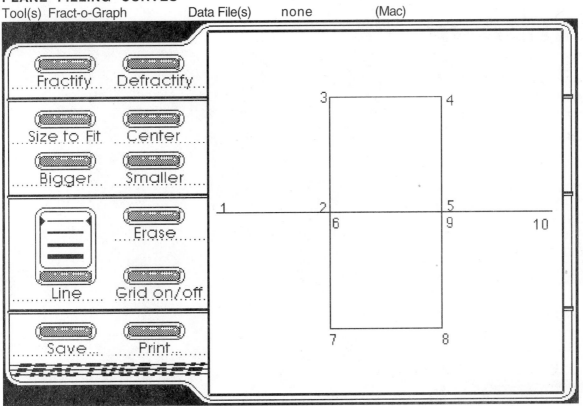

Focus

Some fractal curves have as their limit points all points within some portion of the plane. Such curves are called "plane filling".

Tasks

1. Start Fract-o-Graph and draw the curve shown above, moving from vertex to vertex as indicated by the numbers. Size the curve to fit the screen, then click Fractify to obtain the curves shown in Figures 5.5 through 5.7. Experiment with other starting shapes as the basis for plane filling curves.

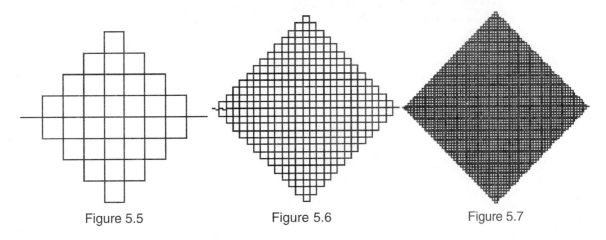

Figure 5.5   Figure 5.6   Figure 5.7

2. Create a fractal curve that resembles a coastline. Save the file for later use.

INTRODUCTION TO FRACTAL DIMENSION

Tool(s) Fractal Coastline   Data File(s)   none   (Mac & JAVA)

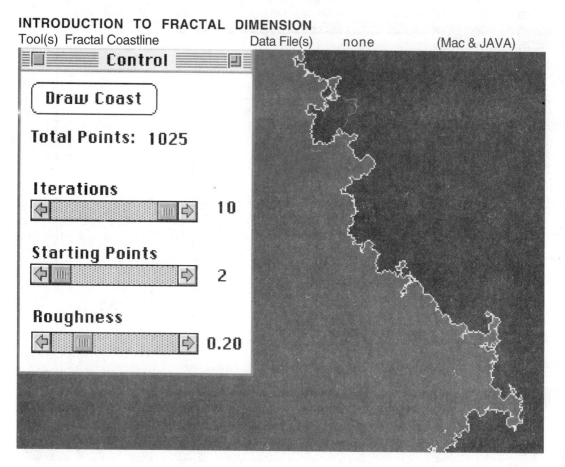

Focus

"How long is the coastline of England?" At first glance, this question seems conceptually simple. It turns out to be complicated. As shorter and shorter "rulers" are used to measure the coastline on a map, the rulers find their way into bays and around peninsulas bridged by longer rulers. Repeated measurements of the coastline using a sequence of shrinking rulers leads to a sequence of lengthening coastlines. Which of these measurements is the "real" coastline of England? This is the question that Benoit Mandelbrot asked himself in the 1950s. He concluded that there is no "correct" answer to this question. Mandelbrot asked a related question, however, that could be answered, "How complex a curve is the coastline of England?" Mandelbrot answered this question by introducing the concept of fractal dimension. Fractal dimension is a number greater than 0 that characterizes a path's complexity. The more complex the path, the higher the fractal dimension of the path.

Fractal dimension is related to conventional concepts of dimensionality. For instance the Euclidean dimension of a line is 1 and the fractal dimension of a line is also 1. Similarly, the Euclidean dimension and fractal dimension of a filled-in square are 2. Also, the Euclidean dimension and fractal dimension of a solid cube are 3. Mandelbrot sought a method for determining the fractal dimension of objects like those shown in Figures 5.1 through 5.7. Since Figure 5.7 is a representation of a plane filling curve, the fractal dimension should be 2, the same as a filled-in square. The method he devised achieves this purpose. In this activity, you will create a coastline and determine its fractal dimension.

Tasks

1. Start Fractal Coastline and set the sliders at 10 iterations, 2 starting points, and a roughness of 0.20. Every time you select Draw Coastline, you will obtain a new coastline. Do so until you obtain a coastline that seems interesting.

2. You will use two methods to determine the fractal dimension of your coastline. The first method measures the length of the coastline using rulers of different length (see Figure 5.8). The second method covers the coastline using grids of different sizes (see Figure 5.9). The results obtained should be similar but not necessarily identical given the limitations of the tool itself.

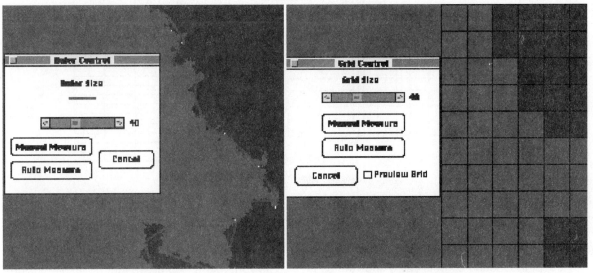

Figure 5.8   Figure 5.9

3. Select Ruler in the Measure pull-down menu. A Ruler Control window will appear. Set the ruler length to 50 units and Auto Measure the coastline. When the Ruler Control window reappears, decrease the size of the ruler to 45 units and Auto Measure again. Repeat this procedure over and over until you have collected at least 8 measurements.

111

4. In the Options pull-down menu, select Data and Show Data. A table will appear containing your measurements. In the Options pull-down menu select Data and Graph Data. Check ruler data and Y vs. X (log-log scale). Click the Curve Fit button to get the equation of the curve. The exponent is the negative of the fractal dimension of the coastline (see Figure 5.10). What is the fractal dimension of your coastline?

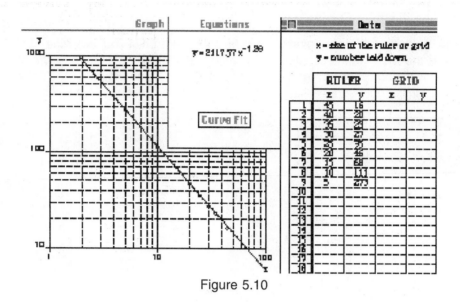

Figure 5.10

5. The procedure for measuring the fractal dimension of the coastline using the grid method is similar. Repeat the measurements using grids the same size as your rulers. Compare the fractal dimensions obtained. Remember to select Grid Data when graphing.

6. Discuss the two methods. Which makes more sense to you? Which is more versatile?

7. Discuss possible uses of fractal dimension other than coastline measurement.

8. Discuss the limitations of Fractal Coastline as a general purpose tool for exploring fractal dimension. What additional features would you like available?

112

MEASURING FRACTAL DIMENSION

Tool(s) Fractal Dimension   Data File(s)   none   (Mac)

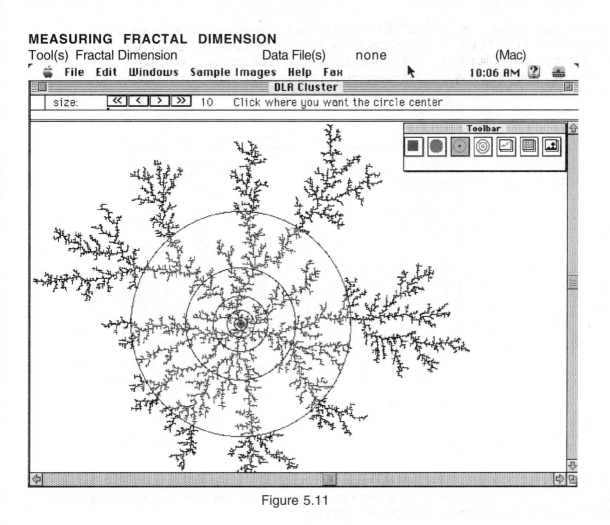

Figure 5.11

Focus

Many things found in nature have fractal properties: coastlines, cracks, lightning, rocks and minerals to name a few. Assuming that you begin with a photograph or image that contains a fractal, the program Fractal Dimension may be used to approximate its fractal dimension.

Tasks

1. Start Fractal Dimension and select the DNA Cluster from the Sample Images pull-down menu.

2. Read the Basic Operation, Data Window, and Tool Bar Info files in the Help pull-down menu.

3. Click "Do fast circle measurement" in the Tool Bar menu. Describe what happens.

4. Click "Bring data window to front" in the Tool Bar menu. How is the data presented?

5. Click "Bring graph window to front" in the Tool Bar menu. How is the graph presented? What is the fractal dimension of the DNA Cluster?

6. Select three other Sample Images and repeat these procedures.

7. You may also use Fractal Dimension to measure the fractal dimension of objects created using other programs or scanned in from photographs. Find the fractal dimension of the coastline you created with Fract-o-Graph. Select Open from the File pull-down menu and find your fractal coastline. You may then measure its fractal dimension.

8. Discuss potential applications of this technology in remote sensing and image processing.

Tool(s) Fract   Data File(s)   none   (Mac)

 File   Edit   Display   Hints   Fax   3:51 F

			Sierpinski Triangle			
	A	B	C	D	E	F	W
1	0.50	0	0	0.50	0	0	0.329
2	0.50	0	0	0.50	1	0	0.328
3	0.50	0	0	0.50	0.50	0.50	0.339

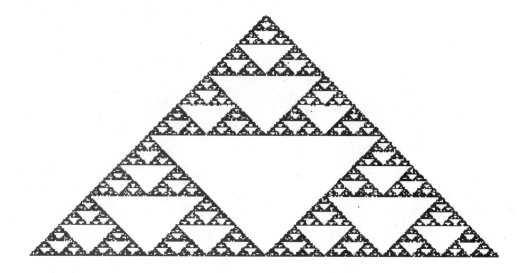

Focus

Linear transformations of the Euclidean plane may be represented in matrix notation as

$$\begin{bmatrix} a & b & e \\ c & d & f \\ 0 & 0 & 1 \end{bmatrix} \cdot \begin{bmatrix} x \\ y \\ 1 \end{bmatrix} = \begin{bmatrix} x' \\ y' \\ 1 \end{bmatrix}$$, where the transformation matrix $\begin{bmatrix} a & b & e \\ c & d & f \\ 0 & 0 & 1 \end{bmatrix}$ is multiplied times object

points $\begin{bmatrix} x \\ y \\ 1 \end{bmatrix}$ to obtain image points $\begin{bmatrix} x' \\ y' \\ 1 \end{bmatrix}$.

For most students, the study of transformation geometry is limited to a review of translations, rotations, reflections, glide reflections, and dilations and their representations in matrix form. For instance, all

translations are of the form $\begin{bmatrix} 1 & 0 & e \\ 0 & 1 & f \\ 0 & 0 & 1 \end{bmatrix} \begin{bmatrix} x \\ y \\ 1 \end{bmatrix} = \begin{bmatrix} x' \\ y' \\ 1 \end{bmatrix}$.

Dilations (uniform rescalings) about the origin are of the form $\begin{bmatrix} r & 0 & 0 \\ 0 & r & 0 \\ 0 & 0 & 1 \end{bmatrix} \begin{bmatrix} x \\ y \\ 1 \end{bmatrix} = \begin{bmatrix} x' \\ y' \\ 1 \end{bmatrix}$, and so on.

Questions about iterated functions, such as "What happens when the same linear transformation is repeated over and over," rarely appear in traditional geometry texts or courses. In general, iterated function systems are not considered at all. Fractal geometry provides a genuine motivation for considering such systems and rewarding visual results.

Tasks

1. Start Fract and Open the file Sierpinski Triangle. The following table of data will appear.

	A	B	C	D	E	F	W
1	0.50	0	0	0.50	0	0	0.329
2	0.50	0	0	0.50	1	0	0.328
3	0.50	0	0	0.50	0.50	0.50	0.339

Each line in the table contains the entries for a 3 x 3 transformation matrix of the form $\begin{bmatrix} a & b & e \\ c & d & f \\ 0 & 0 & 1 \end{bmatrix}$.

Write three transformation matrices, T1, T2, and T3, corresponding to the rows in the table. Ignore the column headed W.

2. Find the image of a unit square with corner points (0,0), (1,0), (1,1), (0,1) under each transformation. Sketch the original unit square and its image under each transformation on the grid below.

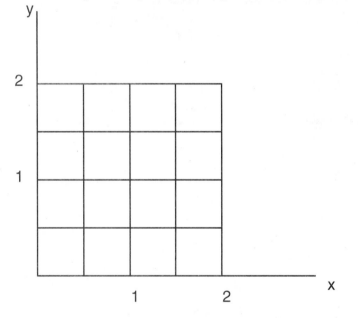

3. Compare the three transformation matrices to the general representations for

translations $\begin{bmatrix} 1 & 0 & e \\ 0 & 1 & f \\ 0 & 0 & 1 \end{bmatrix} \cdot \begin{bmatrix} x \\ y \\ 1 \end{bmatrix} = \begin{bmatrix} x' \\ y' \\ 1 \end{bmatrix}$ and dilations $\begin{bmatrix} r & 0 & 0 \\ 0 & r & 0 \\ 0 & 0 & 1 \end{bmatrix} \cdot \begin{bmatrix} x \\ y \\ 1 \end{bmatrix} = \begin{bmatrix} x' \\ y' \\ 1 \end{bmatrix}$.

Each transformation is a dilation, reducing both the x and y coordinates by half. The transformation in row 1 does not include a translation. The other two transformations do involve a shift in the x and/or y direction. Write a description of each transformation in your own words.

116

4. Fract selects one of the corner points of the original square and finds its images under one of the three linear transformations, selected randomly. The image of that point is then computed in the same manner. After this process is repeated hundreds of times, an image begins to emerge called the "strange attractor" of the iterated function system. To see this happen, Select Draw in the Display pull-down menu. After the image has been plotted in detail, click on the screen, and a grid will appear. You can remove the grid by selecting Grid Off in the Display pull-down menu. Discuss the relationship between the strange attractor and the graph plotted by hand.

5. Select the Mutate option in the Display pull-down menu. Several variations on the Sierpinski triangle will appear. Click on one, then select Editing in the Display pull-down menu. Relate the changes in the table to the changes in the image.

6. Repeat these procedures with the Fern file.

7. Try modifying the Fern file so that it generates a strange attractor resembling a bush or tree. Write the linear transformations below.

Tool(s) Netscape Navigator   Data File(s) none   (Mac & PC)

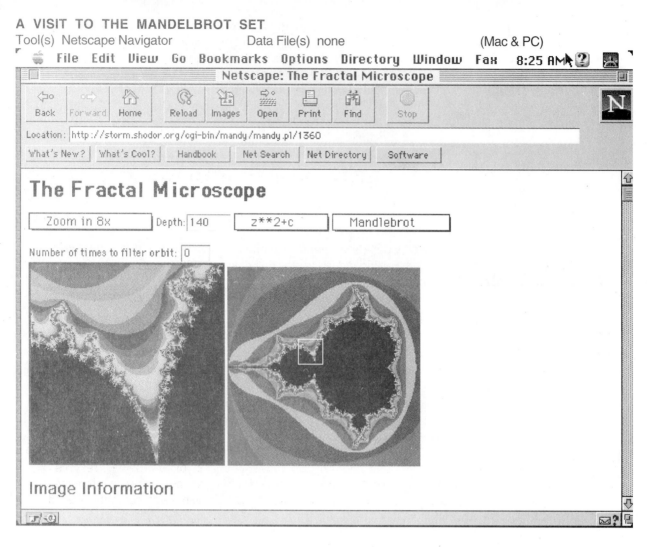

Focus

The Mandelbrot Set may be the most famous object in 20th century mathematics. This activity uses the information and computational resources of the WWW to explore this topic.

Tasks

1. Open the URL http://www.shodor.org/MASTER/fractal_home.html. Take the links to "The Fractal Microscope: an Introduction" and "The Fractal Microscope Homepage at NCSA." Study the materials found at those locations.

2. Take the link to " A Netscape-Accessible Version of the Fractal Microscope" and begin exploring the Mandelbrot Set. Set the zoom factor, then click on the location in the image that you want to explore. An enlargement of that region will appear in the same space.

3. Examine the image archives and submit an image of your own.

CHAPTER 6

MATHEMATICS AND IMAGE PROCESSING

Measuring the Earth and other planets using remote sensing and image processing is an important feature of modern science and technology. This chapter explores the mathematical and technological nature of this sort of activity using a powerful image display and analysis tool called NIH Image. Developed by the National Institutes of Health and implemented in a PC environment by Scion, Inc., NIH Image is a true scientific tool, used by researchers around the world.

The authors of NIH Image describe the program as follows:

NIH Image is a public domain image processing and analysis program for the Macintosh. It can acquire, display, edit, enhance, analyze, and animate images. It supports many standard image processing functions, including contrast enhancement, density profiling, smoothing, sharpening, edge detection, median filtering, and spatial convolution with user-defined kernels.

Image can be used to measure area, mean, centroid, perimeter, and so on of user-defined regions of interest. It also performs automated particle analysis and provides tools for measuring path lengths and angles. Spatial calibration is supported to provide real world area and length measurements. Density calibration can be done against radiation or optical density standards using user-specified units. Results can be printed, exported to text files, or copied to the Clipboard.

A tool palette supports editing of color and gray scale images, including the ability to draw lines, rectangles and text. It can flip, rotate, invert and scale selections. It supports multiple windows and eight levels of magnification. All editing, filtering, and measurement functions operate at any level of magnification and are undoable.

Image can be customized via a built-in Pascal-like macro language. Example macros, plug-ins and complete source code are available from the NIH Image Web site (http://rsb.info.nih.gov/nih-image/) or by anonymous file transfer (FTP) from zippy.nimh.nih.gov.

The images of Mars and Earth used in the activities in this chapter come from NASA, one of the world's great sources of remote sensing data. Using NIH Image and NASA data, students and teachers can explore the Earth and other worlds as scientists do. The activities in this chapter integrate science, mathematics, and technology as students investigate Olympus Mons, the largest volcano in the solar system, global sea surface temperatures on Earth, and other topics.

A VISIT TO OLYMPUS MONS

Tool(s) NIH or Scion Image   Data File(s) nih1 & nih2   (Mac & PC)

Focus

In this activity you will investigate one of Mars' most spectacular features, the shield volcano Olympus Mons. Located at 133 degrees West Longitude and 20 degrees North Latitude, Olympus Mons rises to an elevation of approximately 24,000 m, far above the thin Martian atmosphere. By contrast, Mt. Everest and the Hawaiian volcano Kilauea (measured from its base at the sea floor) are only about 9,000 m in height. The largest known volcano in the solar system, Olympus Mons is a true giant.

Tasks

1. Start NIH Image or Scion Image. When the program is loaded, move the pointer to the File pull-down menu in the upper left-hand corner of the screen, depress the mouse button, drag down to Open, and release the button. Double click on the file nih1. After a few seconds the image shown above will appear. What is your first impression of the planet Mars? How would you describe Mars to someone who has never seen it?

2. Select Close from the File pull-down menu and Open the file nih2. After a few seconds the image shown below will appear. What is your first impression of Olympus Mons? How would you describe it to someone who has never seen it?

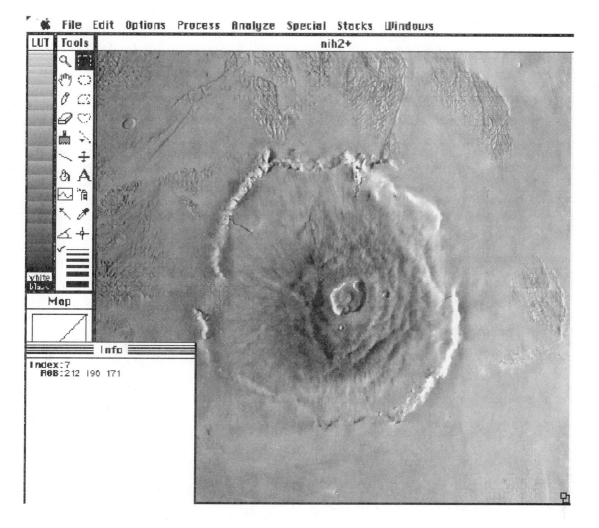

3. Every satellite image is actually made up of small picture elements called "pixels." Click on the magnifying glass in the upper left-hand corner of the Tools window and move it onto the image. Click the mouse button repeatedly until you can see the shape of the pixels. Sketch that shape below. Undo the magnification by double clicking on the magnifying glass in the Tools window.

4. Each pixel in the Olympus Mons image is 1850 m on a side. How many kilometers is that?

5. Suppose that a line on the image is 100 pixels long. How many kilometers is that?

6. Move the pointer to the Analyze pull-down menu at the top of the screen, depress the mouse button, drag down to Set Scale, and let go. A new window will open similar to Figure 6.1. In the center of the window is a box labeled Units. Click on the down arrow in the box and select kilometers as your unit of measurement. At the top of the window enter 1 in the Measured Distance box and 1.850 in the Known Distance box, as shown below. Click OK once you have done so. Now you are ready to measure Mars!

Measured Distance: `0.00` Pixels

Known Distance: `0.00`

Pixel Aspect Ratio: `1.0000`

Units: `Kilometers` ▼

Scale: `0.54054` pixels per `km`

`Cancel`   `OK`

Figure 6.1

7. Select Options in the Analyze pull-down menu and click the Measurement Options shown in Figure 6.2. When you have done so, click OK.

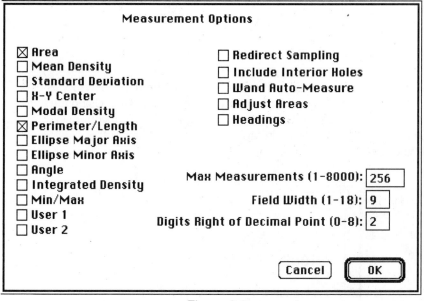

Measurement Options

☒ Area
☐ Mean Density
☐ Standard Deviation
☐ X-Y Center
☐ Modal Density
☒ Perimeter/Length
☐ Ellipse Major Axis
☐ Ellipse Minor Axis
☐ Angle
☐ Integrated Density
☐ Min/Max
☐ User 1
☐ User 2

☐ Redirect Sampling
☐ Include Interior Holes
☐ Wand Auto-Measure
☐ Adjust Areas
☐ Headings

Max Measurements (1-8000): 256
Field Width (1-18): 9
Digits Right of Decimal Point (0-8): 2

[Cancel] [OK]

Figure 6.2

8. Using the Segment Tool (Tools window, 5th icon from the top in the right-hand column), draw a segment across the entire volcanic shield. Once the segment is drawn, return to the Analyze pull-down menu, drag down to Measure, and release the mouse button. Again, click on Analyze, drag down to Show Results, and release the mouse button. A Results window will appear with your answer in the column marked "length." What is the longest such segment you can draw? The shortest? Draw and measure a segment that you think is representative of the diameter of the volcano.

9. Use the magnifying glass in the Tools window to zoom in on the caldera of the volcano. Use the Segment Tool to take a series of measurements of the diameter of the caldera. Which measurement is the most reasonable estimate of the diameter of the caldera? Explain why you think that particular measurement is the best.

10. Using a calculator, determine the area of each pixel in square kilometers (recall that the image is made of pixels 1.850 km on a side). Is each pixel more than or less than 1 square kilometer?

11. Use your answer to question number10 and your calculator to compute the area of each pixel in square meters.

12. NIH Image has four tools for selecting regions for study. In the right-hand column, the first tool is used to drag out rectangular areas. The second tool is used to define oval or circular areas. Using the third tool, you can connect a sequence of points with line segments, creating a polygon of your own design. To close the polygon, just click the mouse twice in rapid succession. The fourth tool allows you to freehand draw any curve you want.

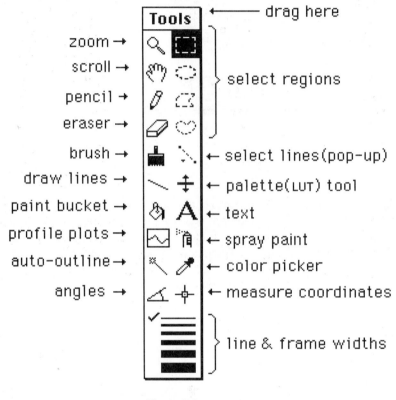

Figure 6.3

Approximate the area of the caldera using each selection tool as accurately as possible.

After making each selection, first Measure then Show Results from the Analyze pull-down menu.

Record your measurements below:

- using the rectangle tool

- using the oval tool

- using the polygon tool

- using the freehand tool

Which measurement do you think is most accurate? Why?

124

13. Approximate the area of the entire volcano using each selection tool.

    After making each selection, first Measure then Show Results from the Analyze pull-down menu.

    Record your measurements below:

    - using the rectangle tool

    - using the oval tool

    - using the polygon tool

    - using the freehand tool

    Which measurement do you think is most accurate? Why?

14. How big is Olympus Mons compared to the state in which you live?

15. Reload the file nih1. The giant canyon cutting across the planet just south of the equator is called Valles Marineris. The scale for that image is approximately 7.4 km per pixel. Enter that figure in the Set Scale option in the Analyze pull-down menu. Take a series of measurements of Valles Marineris. Compare Valles Marineris to the Grand Canyon on Earth. Which is longer? Wider?

MEASURING GLOBAL SEA SURFACE TEMPERATURES

Tool(s) NIH or Scion Image   Data File(s) nih3, nih4, nih4.1, nih5   (Mac & PC)

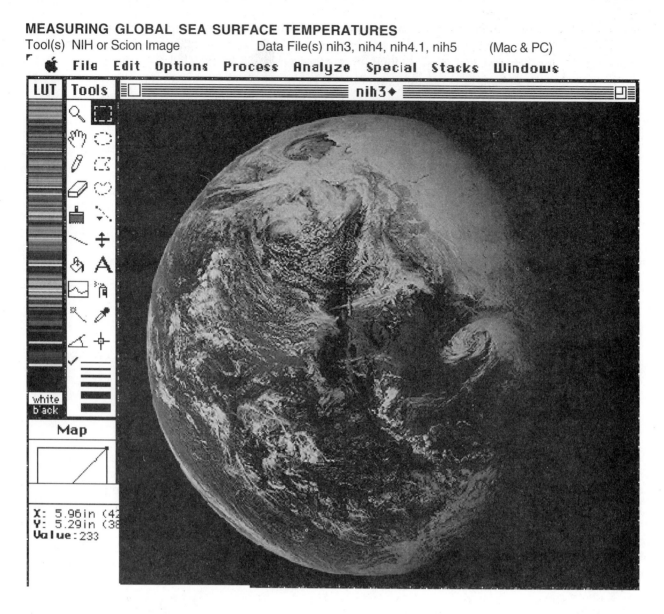

Focus

Water covers approximately 80% of the Earth's surface. The oceans in particular play a major role in driving the Earth's weather systems. Long-term patterns in the weather, called climate, interact with other environmental factors to make one place a lush rain forest and another a desert. These patterns may seem eternal in nature, but they are not. In the past, catastrophic changes in climate have had profound impacts on both sea life and land-based life. For instance, one theory holds that the dinosaurs died from starvation following an asteroid or comet impact that darkened the sky for years with ash and dust, killing the plants that the dinosaurs used for food.

The Ice Ages offer a different lesson in global climate change... it can happen slowly. Some scientists are concerned that the Earth may be about to enter another period of significant climatic change. Others believe that the changes we are now seeing are normal and should not be seen as threatening. In truth, it's too early to tell. But it's not too early to start understanding the scientific issues involved. This activity will open a window on one aspect of this problem, global sea surface temperature, and suggest ways that you can find out more about the engine that drives ourweather, the sea.

The oceans have been monitored for the past 17 years using remote sensing satellites that measure the temperature of the sea very accurately from space. Current satellites do this to within .3 degrees Celsius of the actual sea surface temperature. Accuracy is important because it is believed that small changes in the temperature of the sea can have significant impacts on world-wide weather patterns. The scientific data that you will use was collected by the Advanced Very High Resolution Radiometer (AVHRR) on board the NOAA 7 (National Oceanic and Atmospheric Administration) satellite during the period 1981-1986. Using a scientific visualization tool called NIH Image, your goal is to display and make sense of a small part of this data and to communicate your findings to one another and your teacher.

Tasks

1. Start NIH Image by double clicking on its icon. Select Open from the File pull-down menu. When the dialog box appears, double click on the file nih3. This is the Earth seen from space. If you were an alien visitor, what would your first impression be?

2. Select Close from the File pull-down menu to erase the image of Earth. Select Open from the File pull-down menu. When the dialog box appears, double click on the file nih4. The image that appears represents the average daytime sea surface temperatures during the month of December, 1986. In this image, water temperature is color-coded. The land is colored black. Which colors do you associate with cold water? With warm water?

3. The location of each pixel (picture element) is also coded. Your first task is to understand the location code. Select the cross-hairs icon in the Tools window near the bottom of the right-hand column. Position the cross-hairs near the lower left-hand corner of the image and observe what happens to the x-coordinate and y-coordinate in the Info window as you move the cross-hairs around the image. Find the (x,y) coordinates of each corner of the image.

   - Lower left corner

   - Upper left corner

   - Lower-right corner

   - Upper-right corner

4. What are the dimensions of the image?

   - Rows (horizontal)

   - Columns (vertical)

5. How many degrees of longitude per pixel?

6. How many degrees of latitude per pixel?

7. Do all the pixels represent equal-angle sections of the Earth?

8. Do all the pixels represent equal-area sections of the Earth?

9. In order to understand the relationship between a pixel's coordinates in the image and its corresponding longitude and latitude, it helps to rearrange the image as shown in Figure 6.4. The vertical white line is a line of longitude called the prime meridian. What is the name of the horizontal white line?

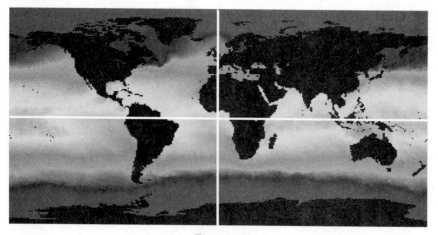

Figure 6.4

10. As shown in Figure 6.5, the pixels may be thought of as equal-angle tiles covering the Earth. These tiles spread out from the prime meridian and equator. The lines of longitude and latitude that name each tile are marked with heavy lines.

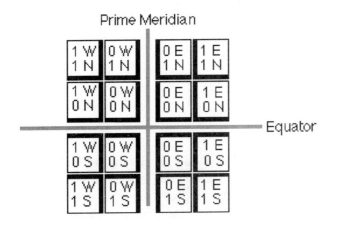

Figure 6.5

11. Figure 6.6 gives the rules for converting the (x,y) coordinates in nih3 to longitude and latitude. This is necessary since most satellite images are not centered on the intersection of the prime meridian and the equator.

	Y <= 89		Y > 89	
X <= 179	[0 ... 179) East X East 89 – Y South	0 ... 89) S	[0 ... 179) East X East Y – 90 North	N (89 ... 0]
X > 179	(179 ... 0] West 359 – X West 89 – Y South	0 ... 89) S	(179 ... 0] West 359 – X West Y – 90 North	N (89 ... 0]

Figure 6.6

Find the following cities using the cross-hairs tool and the (x,y) coordinates. Use Figure 6.6 to convert the cities' (x,y) coordinates to longitude and latitude:

(X,Y)	City	Longitude	Latitude
(104,90)	Singapore		
(302,54)	Buenos Aires		
(116,57)	Perth		

12. The Value of each pixel, shown in the lower left Values window, is called the data number or DN. You can convert the DN (Value) to degrees Celsius using the formula $C = 0.15*DN - 2.1$. Zoom in on each location and find the DN (Value) of the sea pixel closest to each city (black pixels represent land area). Use the DN and formula to find the Celsius temperature.

(X,Y)	City	Degrees Celsius
(104,90)	Singapore	
(302,54)	Buenos Aires	
(116,57)	Perth	

13. Select Load Macros in the Special pull-down menu. When the dialog box opens, Open the file nih4.1. A new option labeled Temperature will be available in the Special pull-down menu. Select Temperature and observe the data displayed in the Info box in the lower left-hand corner of the screen as you move the cursor around the image. What is the lowest temperature in Fahrenheit degrees in the image? The highest temperature in Fahrenheit degrees?

14. By now you should have some idea what the colors mean in the global sea surface temperature image. Select the cross-hairs icon and move the cursor into the LUT (Look Up Table) window. As you move the marker from the top (purple) to the bottom (red) of the color table, the DNs (called Index in the Values window) change from 0 to 254. Think of this color scale as a key to the temperatures in the image, where cold things are purple and warm things are red. Colors in between purple and red represent temperatures in between cold and hot. Note that the land is black, and its DN is 255, regardless of the temperature. What temperature is associated with a DN of 0? A DN of 254?

15. Select the Rectangle Tool from the Tools window. Use it to draw a rectangle starting at (155,89) and extending to (234,137). Select Load Macros from the Special pull-down menu. When the dialog box appears, select the Macros folder. Select the macro entitled Plotting Macros. Pull-down the Special menu again and select Plot Histogram.

The histogram shows the relative frequency of various colors (temperatures) in the rectangle. The most frequent color has the greatest height in the histogram. Click on the histogram and position the cross-hairs on the highest spot.

What is the most frequent DN (Value) in your rectangle?
Convert this DN to degrees Celsius.

What is the highest DN (Value) in your rectangle?
Convert it to degrees Celsius.

What is the lowest DN (Value) in your rectangle?
Convert it to degrees Celsius.

16. Write a few sentences in which you interpret the general shape of the histogram in terms of the sea temperature within the rectangle.

17. Click on the histogram. In the upper left-hand corner is a small box. Click the box. You will be asked if you wish to save changes in Histogram. Click Yes. A dialog box will appear with Histogram highlighted in the Save As box. Give a name to your histogram, such as North Pacific Histogram and click Save. This will save your histogram for later use.

18. You should still see your rectangle in the image. If not, click on the image and it should reappear. Select the Hand tool from the Tools window, move the hand into your rectangle, and drag it straight down so that the top edge of the rectangle lies on the equator and the left-hand edge of the rectangle is on the point (155,89). Pull down the Special menu and click on Plot Histogram to get a histogram of the South Pacific. Drag the histogram off the image so that you can see both clearly.

What is the most frequent DN (Value) and temperature in your rectangle?

What is the highest DN (Value) and temperature?

What is the lowest DN (Value) and temperature?

Write a few sentences interpreting the histogram.

19. As before, save the histogram and name it South Pacific Histogram. Pull down the File menu and click on Open. When the dialog box appears, select North Pacific Histogram. Pull down the File menu again and Open South Pacific Histogram. Move the two histograms so that their edges line up, permitting an easy comparison. Write a few sentences comparing the two histograms and interpreting their differences in terms of water temperature.

20. Pull down the Enhance menu and select Image Math. When the dialog box appears, select North Pacific Histogram - South Pacific Hemisphere and click OK. Write a few sentences describing and interpreting the result.

21. Pull down the File menu and Close all the open files. Pull down the File menu and Open nih5. This image represents global sea surface temperatures during the month of June, 1986. What comparisons could you make using the data from the month of December and the month of June in 1986? What differences would you expect to find? Conduct your investigation and summarize your findings in a paragraph or two.

DETERMINING THE FRACTAL DIMENSION OF AN OBJECT USING NIH IMAGE

Tool(s) NIH or Scion Image   Data File(s) nih6, nih6.1   (Mac & PC)

	Size	Number
1.	1.00	3288.00
2.	5.00	339.00
3.	10.00	124.00
4.	15.00	76.00
5.	20.00	51.00
6.	25.00	34.00
7.	30.00	28.00
8.	35.00	27.00
9.	40.00	21.00
10.	45.00	20.00
11.	50.00	12.00

Info

X : 277
Y : 188
Value : 0

Count : 11
Area : 0 square pixels
Mean : 0.00
Size : 50.00
Number : 12.00

Focus

An NIH Image macro may be used to generate data for estimating the fractal dimension of a curve. When the data is plotted on log-log paper, the fractal dimension is associated with the slope of the "best fit" straight line through the data points.

Tasks

1. Select Open in the File pull-down menu and double click on nih6. In the Special pull-down menu, select Load Macros. When the dialog box appears Open the file nih6.1. A new option, Fractal Dimension, will appear in the Special pull-down menu. Select Fractal Dimension and wait for the Results window to appear. In the File pull-down menu, select Save As. When the dialog box appears, click the Text option and Save the file as nih6(Text). Print the file from the File pull-down menu.

2. Plot the data points on log-log graph paper. Draw a straight line through the data points representing the "best fit." Select two points on the line, (X1,Y1) and (X2,Y2), and compute the slope using the formula: Fractal Dimension = abs (log Y1 − logY2) / (log X1 − log X2).
   *Note: Using **ln** instead of **log** will produce the same answer.*

3. Using a graphing calculator or spreadsheet,
   - Copy the data into your calculator or spreadsheet program.
   - Find the log of every data value.
   - Perform a linear regression on the new log-log data. The slope of the line is the fractal dimension.

MAP COLORING

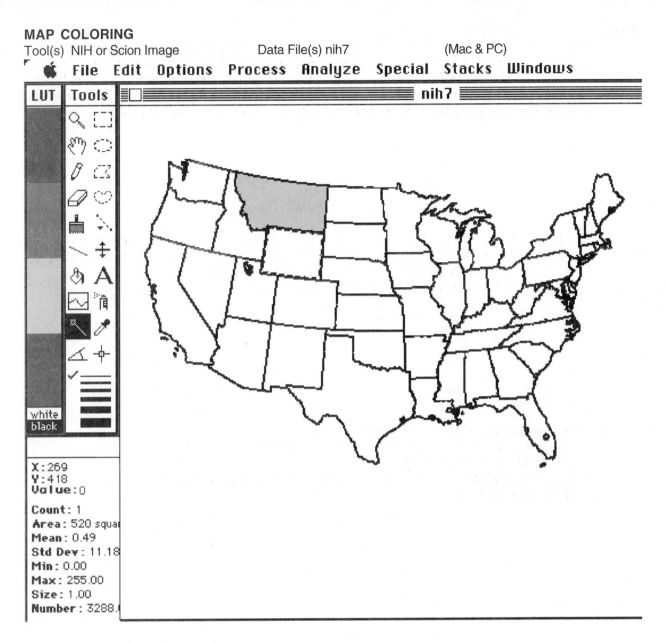

Focus

When maps are colored, adjacent regions must be different colors. What is the minimum number of colors needed to shade any map in this manner? NIH Image is used to explore this question.

Tasks

1. Select Open in the File pull-down menu. When the dialog box appears, open the file nih7. Using the Auto Outline tool (highlighted in the Tools window above), click the mouse in Montana. What happens to the state boundary?

2. Select the Color Picker in the Tools window and move it into the LUT window. Click on one of the colors there. Select Fill in the Edit pull-down menu. Describe what happens.

3. Working outward from Montana, color as many states as possible using just two colors. Which state boundaries cannot be colored using only two colors?

4. Repeat the process using three colors.

5. Repeat the process using four colors.

6. Under the File pull-down menu, select New. When the dialog box appears, specify window dimensions of 500x300 pixels. When the window appears, use the Color Picker tool to select the color black and the Pencil tool to draw a map such as that shown in Figure 6.7. Color your map with as few colors as possible.   Create maps that require exactly two colors, three colors, four colors. Try to create a map that requires 5 colors.

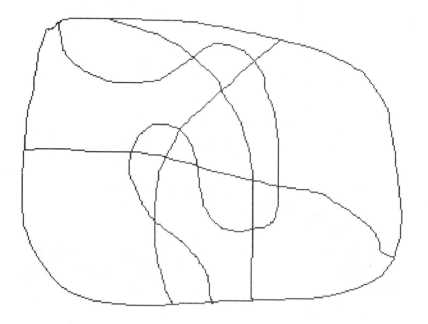

Figure 6.7

INSIDE OR OUTSIDE?

Tool(s) Geometers Sketchpad   Data File(s) gsp64, gsp65   (Mac & PC)

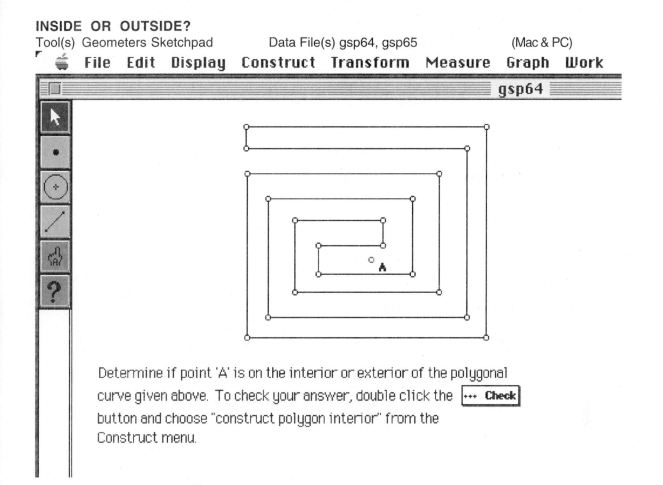

Determine if point 'A' is on the interior or exterior of the polygonal curve given above. To check your answer, double click the ⋯ **Check** button and choose "construct polygon interior" from the Construct menu.

Focus

Is point A in the interior of the curve or is it exterior?

Tasks

1. Study the location of point A in gsp64. How could you determine whether point A is in the interior of the curve? Double click on the Check button to see a demonstration. Describe the strategy used.

2. Open the file gsp65 and study the location of point B. Double click on the Check button to see a demonstration. Describe the strategy used.

3. Describe a different strategy than the ones used in gsp64 and gsp65.

Suggested Readings

Banchoff, T. F. (1990). Beyond the third dimension: Geometry, computer graphics, and higher dimensions; Scientific American Library. New York: W. H. Freeman.

Barnsley, M. (1988). Fractals everywhere. San Diego: Academic Press, Inc.

Conway, J., Doyle, P., & Thurston, W. Geometry and the imagination. Research report GCG27. The Geometry Center, 1300 So. 2nd St., Minneapolis, MN 55454.

Dewdney, A. K. (1988). The armchair universe. New York: W. H. Freeman.

Gleick, J. (1987). Chaos: Making a new science. New York: Viking.

Mandelbrot, B. (1982). The Fractal Geometry of Nature. New York: W.H. Freeman and Co.

Mandelbrot, B. (1977). Fractals: Form, chance, and dimension. San Francisco: W. H. Freeman.

Norman, J. & Stahl, S. (1979). The mathematics of Islamic art. New York: The Metropolitan Museum of Art.

Peitgen, H. O., Jurgens, H., & Saupe, D. (1992). Fractals for the classroom, part two. New York: Springer-Verlag.

Peitgen, H. O., Jurgens, H. & Saupe, D. (1991). Fractals for the classroom, part one. New York: Springer-Verlag.

Peitgen, H. O. & Saupe, D. (1988). (Eds.). The science of fractal images. New York: Springer-Verlag.

Peitgen, H. O., Jurgens, H., Saupe, D., Maletsky, E., Perciante, T., & Yunker, L. (1992). Fractals for the classroom: Strategic activities, volume two. New York: Springer-Verlag.

Peitgen, H. O., Jurgens, H., Saupe, D, Maletsky, E., Perciante, T., & Yunker, L. (1991). Fractals for the classroom: Strategic activities, volume one. New York: Springer-Verlag.

Thomas, D. (Ed.). (1995). Scientific visualization in mathematics and science teaching. Charlottesville, VA: Association for the advancement of computing in education. ISBN 1-880094-09-6.

Thomas, D. (1995). Math projects in the computer age. New York: Franklin-Watts. ISBN 0-531-11213-6.

Thomas, D. (1992). Teenagers, teachers, & mathematics. Needham Heights, MA: Allyn & Bacon. ISBN 0-205-13194-8.

Thomas, D. (1992). Using computer visualization to motivate and support mathematical dialogues. Journal of computers in mathematics and science teaching, 11(3/4), 265 – 274.

Thomas, D. (1991). Children, teachers, & mathematics. Needham Heights, MA: Allyn & Bacon. ISBN 0-205-12681-2.

Thomas, D. (1990). Turtle tessellations. Journal of computers in mathematics and science teaching, 9(4), 25–34.

Thomas, D. (1989). Investigating fractal geometry using LOGO. Journal of computers in mathematics and science teaching, 8(3), 25–31. Reprinted in The Illinois Mathematics Teacher, 40(3) 21–29 ATACC Journal, 8(1), 9–17.

Thomas, D. (1989). Tessellations: A blend of art and mathematics. Virginia MathematicsTeacher, 15(3), 1–3.

Thomas, D. (1988). Math projects for young scientists. New York: Franklin-Watts. ISBN 0-531-10523-7 & ISBN 0-531-15133-6.

Thomas, D. & Thomas, C. (1997). Using Internet-based K-12 classroom activities: Materials & staff development. INET'97: Proceedings of the 1997 meeting of the Internet Society. Kuala Lumpur, Malaysia, 24-27 June, 1997. URL http://www.isoc.org/isoc/whatis/conferences/inet/97/proceedings/D1/D1_2.HTM

Zimmerman, W. & Cunningham, S. (Eds.). Visualization in teaching and learning mathematics. Washington, DC: Mathematical Association of America.

COMPUTER SOFTWARE AND DATA FILES

All computer tools and data files used in this book are maintained on the Brooks/Cole WWW site. Using a web browser such
as Netscape Navigator® or Microsoft Internet Explorer®, open the
URL **http://www.brookscole.com/math/authors/thomasd/**
The following page will appear. If you have never used a WWW
browser to download files, or if the computer you are using does
not have either WinZip® or StuffitExpander® properly installed,
you may need *local* assistance.

Please do not ask the Brooks/Cole systems operator or author to
help you configure your computer or internet connection.

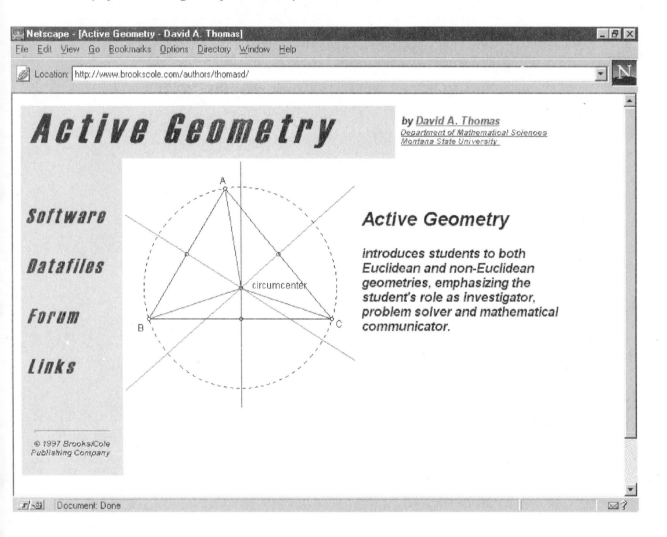